RECENT THEMES IN THE PHILOSOPHY OF SCIENCE

AUSTRALASIAN STUDIES IN HISTORY AND PHILOSOPHY OF SCIENCE

VOLUME 17

RECENT THEMES IN THE PHILOSOPHY OF SCIENCE

Scientific Realism and Commonsense

Edited by

STEVE CLARKE
Charles Sturt University, Canberra, Australia

and

TIMOTHY D. LYONS
Indiana University – Purdue University Indianapolis, Indianapolis, U.S.A.

KLUWER ACADEMIC PUBLISHERS
DORDRECHT / BOSTON / LONDON

A C.I.P. Catalogue record for this book is available from the Library of Congress.

ISBN 1-4020-0831-7

Published by Kluwer Academic Publishers,
P.O. Box 17, 3300 AA Dordrecht, The Netherlands.

Sold and distributed in North, Central and South America
by Kluwer Academic Publishers,
101 Philip Drive, Norwell, MA 02061, U.S.A.

In all other countries, sold and distributed
by Kluwer Academic Publishers,
P.O. Box 322, 3300 AH Dordrecht, The Netherlands.

Printed on acid-free paper

Printed in the Netherlands.

TABLE OF CONTENTS

FOREWORD

Australia and New Zealand boast an active community of scholars working in the field of history, philosophy and social studies of science. *Australasian Studies in History and Philosophy of Science* aims to provide a distinctive publication outlet for their work. Each volume comprises a group of thematically-connected essays edited by scholars based in Australia or New Zealand with special expertise in that particular area. In each volume, a majority of the contributors are from Australia or New Zealand. Contributions from elsewhere are by no means ruled out, however, and are actively encouraged wherever appropriate to the balance of the volume in question. Earlier volumes in the series have been welcomed for significantly advancing the discussion of the topics they have dealt with. I believe that the present volume will be greeted equally enthusiastically by readers in many parts of the world.

R.W. Home
General Editor
Australasian Studies in History
And Philosophy of Science

ACKNOWLEDGEMENTS

The majority of the papers in this collection had their origin in the 2001 Australasian Association for History, Philosophy, and Social Studies of Science annual conference, held at the University of Melbourne, where streams of papers on the themes of scientific realism and commonsense were organised. It soon became apparent that many of the papers in these streams were of unusually high quality and that there was a strong case for organising a new volume of the *Australasian Studies in History and Philosophy of Science* series based on these papers. The collection was further strengthened when Herman De Regt, Harold Kincaid and Keith Hutchison were invited to submit papers relevant to the combined themes of the collection.

All of the papers in this collection have been anonymously refereed and have been considerably revised from their original form. Thanks to Rod Home, the General Editor of the *Australasian Studies* series and to Stephen Gaukroger, who will assume the General Editorship of the series in 2003, for their encouragement and support in having this volume approved for publication. Thanks also to Howard Sankey who has been a constant source of good advice about the editing process. And thanks to Stephen Ames, Alan Chalmers, Wayne Christensen, Brian Ellis, Frank Jackson, Cathy Legg, Steve Matthews, Don Ross, David Spurrett, John Sutton and Neil Thomason for their invaluable assistance.

Steve Clarke

Timothy D. Lyons

TIMOTHY D. LYONS AND STEVE CLARKE

INTRODUCTION: SCIENTIFIC REALISM AND COMMONSENSE

1. SCIENTIFIC REALISM

Scientific realism involves two key claims. First, science aims primarily at truth. Second, we can justifiably believe that our successful scientific theories achieve, or at least approximate, this aim. The contemporary scientific realism debate turns on the acceptability of these claims. To acquire a more robust picture of scientific realism, let us identify some of the related theses on which these key claims rest.

In opposition to, say, solipsists, the scientific realist insists that there exists an 'external' world with which we interact. Contra social constructivists, the scientific realist holds that this world includes events, processes, and/or entities that are not contingent on our beliefs. Scientific realists take truth to be objective and to express a correspondence relation between statements and the world. Such a conception of truth is often juxtaposed against those conceptions espoused by internal realists (e.g., Hilary Putnam, Brian Ellis).[1] Opposing idealists such as Berkeley, the scientific realist maintains further that we can be justified in believing that the objects we observe exist and that our basic claims about their observable properties are true. In contrast to classical instrumentalists, such as Ernst Mach, positivists (e.g., Moritz Schlick, Rudolph Carnap), as well as fictionalists, operationalists, and phenomenalists, the scientific realist construes scientific theories literally; most terms contained in scientific theories are intended to refer to real entities.[2] Scientific realists hold that, in general, theory change in science has been rational and progressive. Moreover, scientific realists tend to espouse the view that progress in science is determined by the extent to which its primary aim is achieved (or approximated).

These tenets of scientific realism collectively serve to provide a framework within which the *contemporary* debate on scientific realism takes place. Most prominent contemporary opponents of scientific realism

S. Clarke and T.D. Lyons (eds.), Recent Themes in the Philosophy of Science, ix-xxiii.

— such as Bas van Fraassen and Larry Laudan — do not criticize this framework. Rather, the *contemporary debate* on scientific realism hinges primarily on the axiological and epistemological claims noted above. These can be made more explicit:

Axiological (Scientific) Realism: science aims, primarily, to express true statements about the world.

Epistemic (Scientific) Realism: we can be justified in believing that successful scientific theories are (approximately) true.

The majority of philosophers involved in the scientific realism debate assume that axiological realism rests on epistemic realism. In fact, so long as we take science to be successful, progressive, and rational, and so long as progress is determined by the achievement of (or the degree to which we approximate) our primary aim, truth, a defence of epistemic realism is required of any scientific realist. For this reason, the contemporary debate on scientific realism is, by and large, played out in the arena of epistemic realism.

So long as we interpret scientific theories literally, as the scientific realist advises, epistemic realism entails the claim that we are justified in believing that unobservable entities postulated by our successful theories exist. The type of inference that scientific realists usually put forward to support such a claim can be expressed as follows: The existence of an unobservable entity, U, (e.g., the electron) is the best explanation for the observable phenomena, O (e.g. observed electrical phenomena); therefore, we are justified in believing that U (e.g. the electron) exists. An argument of this sort is called an *inference to the best explanation* (IBE). It is generally thought to be the mode of inference that grounds or provides justification for epistemic realism.

Although IBE is employed to support our belief in the existence of unobservables, scientific realists maintain that it is not an 'exotic' mode of inference, utilized only by philosophers. They contend that scientists themselves employ IBE. In fact, realists tell us, IBE plays an integral role in our commonsense reasoning. Bas van Fraassen (though a non-realist) provides a nice example:

> I hear scratching in the wall, the patter of little feet at midnight, my cheese disappears — and I infer that a mouse has come to live with me. Not merely that these apparent signs of mousely presence will continue, not merely that all the observable phenomena will be as if there is a mouse; but that there really is a mouse. (van Fraassen 1980, pp. 19-20)

Scientific realists seek to justify belief not merely in the existence of particular entities but in the (aproximate) truth of our scientific theories. Toward this end, they typically apply a robust version of IBE. This is the 'no-miracles argument', made famous by Putnam (1975) — also known as the 'miracle argument,' the 'success argument' and the 'ultimate argument' — if our successful scientific theories were not at least approximately true, then their success would be a miracle. In other words, so long as we do not accept miracles as explanatory,[3] the *only* (and thus the best) explanation for a theory's success is that the world is as the theory says it is. If we accept this argument, we appear to be led to epistemic realism. And since the belief that our theory is (approximately) true entails the belief that the entities postulated by the theory exist, the no-miracles argument justifies the latter in so far as it justifies the former. Thus the no-miracles argument warrants a far greater range of beliefs than would be warranted by any specific inference to the existence of an unobservable entity.

Along with Alan Musgrave (1988), one could consider the no-miracles argument, as stated thus far, to be more akin to a slogan than an argument. Noting this, we are prompted to explicate it more precisely. Scientific realists typically claim that IBE is abductive, abduction being a form of reasoning famously articulated and advocated by C.S. Peirce (1958). Peirce construes abductive reasoning in the following way. We begin with a 'surprising' observation, (Q). A state of affairs is postulated, and that postulate, (P), would render (Q) 'a matter of course'. We conclude that 'we have reason to suspect' that (P) obtains (1958, p. 189).

While scientific realists often tip their hats to Peirce, when presenting the no-miracles argument, the way in which it is to be expressed as a Peircian abduction is neither obvious nor generally explicated. We can begin by inserting the central realist claims into Peirce's argument. The scientific realist wants to direct our attention to the 'surprising fact,' (Q), that we have successful scientific theories. According to scientific realists, if our theories were (approximately) true, (P), then (Q) would be 'a matter of course.' The epistemic realist draws a bolder conclusion than that drawn by Peirce. The epistemic realist infers, not merely that 'we have reason to suspect' (P), but that we *are justified in believing* (P). This extra step might be legitimised if the epistemic realist can show that (P) is probable. But on what grounds does the epistemic realist base such a claim? Namely, her assertion that, aside from (P), the only state of affairs that could bring about (Q) would be a miracle. With this key premise of the no-miracles argument, we are closer to formulating that argument as an abduction. However, at least a few more hidden premises must be made salient.

1: Our theories are successful, (Q)

2: If our theories were (approximately) true, (P), then their success, (Q), would be a matter of course

3: The relationship expressed in (2) shows that the (approximate) truth of our theories, (P), provides an explanation of their success, (Q)

4: In fact, the (approximate) truth of our theories, (P), provides a *good* explanation of success, (Q)

5: To say that success, (Q), occurs due to a miracle is to provide no explanation at all

6: Aside from the (approximate) truth of our theories, (P), there is no other explanation available for their success, (Q)

Therefore, (probably) our theories are (approximately) true, (P)

Therefore, epistemic realism: we are *justified in believing* that our successful theories are (approximately) true, (P).

Though it is a start, this modified abductive argument does not exhaust the list of presuppositions involved in the no-miracles argument. Premises (3), (4), and (6) surely need further clarification and support. And even including our new premises and their requisite support, the full set of premises would *entail* neither the initial, nor the subsequent, conclusion. We are not *logically compelled* to infer from a phenomenon to its explanation, even if that explanation is the only one available. The argument, as a whole, is not deductively valid.[4] The scientific realist will grant this and will remind us that certainty about the world of experience is an unattainable demand. The argument is only meant to ground epistemic realism. It doesn't tells us what we can be certain of, but only what we can be justified in believing. And again, says the scientific realist, it receives its legitimacy from its use, not only in science, but in everyday life.

We have thus far been considering rather basic formulations of epistemic realism and the no-miracles argument. We can now note that scientific realists have introduced a number of variations of this inferential package. Some appeal to *truth* as the explanation of success (e.g., Wilfred Sellars (1962); André Kukla (1997)), while others appeal to *approximate truth* (e.g., Hilary Putnam (1975); Richard Boyd (1973)). Some claim that truth (or approximate truth) is the *only* possible or available explanation for success (e.g., J. J. C. Smart (1968); Putnam (1975)). Others (most contemporary scientific realists) claim the truth or approximate truth of our theories to be the *best* explanation of success. Some scientific realists understand that which is being explained as *the success of a given theory* (e.g., Smart (1968); Musgrave (1985); Peter Lipton (1993), (1994)), while others see that which needs to be explained as being *the success of science in general* (e.g., Putnam (1975)). Some appeal to *general predictive success* (e.g. Putnam (1975); Boyd (1973); W. H. Newton-Smith (1981)), while others emphasize *novel success* (e.g., William Whewell (1840); Musgrave (1985), (1988); Lipton (1993), (1994); Stathis Psillos (1999); Howard Sankey (2001)). Some say we are justified in believing theories as wholes, while others focus on certain constituents of those theories (e.g., Philip Kitcher (1993), Psillos (1999); Sankey (2001)).

It is noteworthy that many scientific realists see their philosophy to be an 'overarching empirical hypothesis.' Scientific realism is taken to be an empirically testable position that shares the virtues of a scientific theory. Acknowledging this, we can clarify epistemic realism further. Epistemic realism is the thesis that we can be justified in believing the *hypothesis* that our successful scientific theories are (approximately) true.

With a framework in hand for understanding the position of scientific realism let us identify three important contemporary objections to that position. One is directed at the no-miracles argument specifically. Some non-realists contend that the scientific realist has put forward a false dichotomy. In seeking an explanation for the success of scientific theories, we need not make a choice between appealing to miracles and inferring that our theories are (approximately) true. Alternative explanations are available (challenging premise 6 above). For example, van Fraassen (1980) presents a Darwinian alternative: success is a requirement for a theory's survival; we simply wouldn't have retained our theories were they not successful. Other alternatives are offered by Laudan (1985), Rescher (1987), Fine (1984), Lyons (this volume), Worrall (1989), and Carrier (1991); (1993). Such alternative explanations deflate the motivation for inferring the epistemic realist hypothesis. They thus cut at the heart of epistemic realism. In reply, the epistemic realist often claims that (approximate) truth provides a *better* explanation than the non-realist contenders. Or she draws attention to a new 'surprising fact' (e.g., to novel success) and denies the non-realist's ability to explain it.

Another non-realist argument is the argument from the underdetermination of theories by data (Duhem (1906), Quine (1975), van Fraassen (1980)). In its basic formulation, this argument proceeds as follows. Any successful theory will have a high (if not infinite) number of empirically equivalent, yet incompatible, rivals. Since each of these rivals will share the empirical success of our preferred theory, we cannot be justified in believing one theory over the others. So, despite the success of our preferred theory, we cannot be justified in believing that it is true. In response, some scientific realists deny that we can generate empirically equivalent rivals for every theory. The debate on this matter continues (see, for instance, Kukla (1997)). Another realist strategy is to argue that empirical success is not the only epistemically relevant virtue. We can still select among our theories by appeal to supra-empirical virtues such as simplicity. Non-realists challenge the assumption that these supra-empirical virtues bear on truth (e.g., van Fraassen, (1980)), thus questioning whether they can be legitimately employed as a justification for belief.

A third important non-realist argument is historical. This argument takes seriously the claim that scientific realism provides an empirically testable hypothesis. It begins with a list of successful theories that cannot, by present lights, be construed as true, or even approximately true. The most commonly discussed version of this argument is known as the pessimistic meta-induction: we have had many successful theories that have now turned out to be false; so our present-day theories will be likely to turn out to be false as well (Putnam (1984); Rescher (1987); Laudan (1977)). A second version involves employing the list, not as fuel for an induction toward the falsity of our theories, but as a set of empirical data that directly counters scientific realism. (See Laudan (1981), as interpreted by Lyons (this volume).)

This edited collection contains a number of papers that engage in contemporary debates about scientific realism in one way or another. Some participants in debates about scientific realism have sought compromises between scientific realism and full-blown antirealism. On possible compromise is to adopt realism about the existence of some of the entities described in scientific theories, without grounding that realism in realism about the truth of scientific theories. This form of compromise is known as entity realism. Entity realism is typically grounded in an epistemology that privileges truth claims that are based on experimental manipulation. Robert Nola considers two explanationist arguments toward such a realism: a variation of the no-miracles argument and an inference to the most probable cause. Nola seeks, however, to extend realism further, arguing that we are justified in believing in the existence of some non-manipulated entities. He contends that, while explanationist arguments may prove inadequate in respect to such entities, there remains a strong probabilistic argument that will, in the appropriate circumstances, lead us to realism about both manipulated and non-manipulated entities. While other entity realists, such as Nancy Cartwright (1983), have considered themselves to be opponents of scientific realism, Nola offers us a means to bridge the gap between entity realism and theory realism. Along with Cartwright (1983), Steve Clarke (2001) defends an entity realism about manipulated entities while rejecting realism about fundamental scientific theories. Here he continues to make a case for an entity realism, which remains unreconciled with scientific realism, defending Nancy Cartwright's account of entity realism in the face of criticisms of it due to Alan Chalmers.

Harold Kincaid argues that both sides of the realist debate employ, at least tacitly, the following assumptions: the rules of scientific inference are global and the justification for these rules is philosophical. He challenges these assumptions by way of an in-depth adjudication between Bayesians and error-statisticians: He contends that both statistical approaches require the case-by-case assessment of empirical factors, the relevance of which has been largely unacknowledged. Drawing on his findings here, he proposes, an alternative form of scientific realism that is contextualised to particular domains and requires domain-specific empirical data.

Timothy Lyons criticises a variety of truth-based formulations of scientific realism. He clarifies the nature of the threatening historical argument against realism and makes explicit the implications that hold at each level of realism. He shows that numerous novel successes have come from false theories, thus that the realist's frequent appeal to novel success does not solve the historical problem. He also considers an alternative explanation for success, and challenges the key explanatory premise of the realist's argument, as it stands once the realist retreats to approximate truth. Strengthening the evidence for the historical argument against realism, Keith Hutchison undertakes an in-depth case study of W. J. M. Rankine's oft neglected, yet surprisingly successful, 19th century vortex theory of heat. Against a sophisticated form of the no-miracles argument, he contends that Rankine's theory contains a number of false theoretical claims that were centrally employed in the derivation of its successful predictions.

Putnam's no-miracles argument, in its full formulation, is carefully unpacked by Michel Ghins. Preferring a philosophically grounded realism, Ghins argues against realists, such as Putnam, who construe their position to be scientifically or naturalistically grounded. Realism does not have the status of a scientific theory; for, among other problems, according to Ghins, truth is not explanatory. John Wright identifies three surprising features of science that seem to call for explanation: the surprising novel successes of our theories; the true descriptions of parts of the world that were, at the time described, unobserved; and the extended empirical success of theories that are initially preferred for their a priori preferable properties. He works his way through a series of proposals that have been, or might be, put forward to explain these phenomena, and he contends that none are suited to the task. Brian Ellis articulates a metaphysics for a specific type of scientific realism, which he calls a causal process entity realism. Along the way, he employs his own version of the no-miracles argument to justify the view that the world has a natural kind structure. Our discussion of the papers in this volume will be continued below.

We have seen above that, underlying the scientific realism debate, are a number of foundational tenets. Some of these are essential to our commonsense conception of reality, for instance, the thesis that we interact with an 'external' world and that our basic beliefs about observable entities are warranted. Also, we have seen that the scientific realist draws on commonsense in her attempt to legitimise abduction, the mode of reasoning on which her position rests. We now attend more directly to commonsense.

2. SCIENCE AND COMMONSENSE

We can understand commonsense as the conceptual framework that ordinary people share. It is the set of, often implicit, assumptions that the majority utilise when they interpret the world and the behaviour of their fellows. We do not believe that everyone has commonsense, but nor do we believe that commonsense is in short supply. Commonsense has two main components.[5] First, it involves a strident form of naive realism. If we see and feel a wooden table in front of us, then it is plain commonsense to believe that a wooden table exists, which we are seeing and feeling. Second, commonsense involves the acceptance and utilisation of the psychology of rational agency, both for self-description and for the interpretation of the behaviour of others. It is commonsensical to believe that people's behaviour is, in the main, goal oriented and that the structure of an individual's beliefs and desires can be appealed to in order to explain why they do what they do. The belief that newspapers can be purchased at newsagencies and the desire to have a newspaper are appealed to by commonsense to explain why a person went to a newsagency and bought a newspaper.

Because commonsense involves a commitment to the mind-independent reality of objects in the world it is most directly opposed to skeptical and idealist antirealisms. The best known advocates of commonsense in the history of philosophy, Thomas Reid (1710-1796) and G. E. Moore (1873-1958), are also mostly concerned to oppose idealist and skeptical arguments. Reid and his followers, the Scottish Commonsense School, develop arguments to buttress the metaphysical stance of commonsense, and to expose the weaknesses of Humean skepticism and Berkeleyan idealism. Moore, who was most directly opposed to the idealism of Bradley and other

British idealists, did not so much elaborate the naive realism of commonsense, as exemplify the force of appeals to commonsense and the concomitant weakness of philosophical arguments that opposed commonsense. Famously, Moore's method of refuting skepticism was to hold up his hand and say 'here is a hand'. In effect, Moore argued that we are more confident that we are in possession of hands, when we see and feel them, than we could be confident of the force of any possible philosophical argument that defies our commonsensical judgement that we have hands.

Science conceptualises the ordinary objects and agents of the world very differently from commonsense. Where commonsense assures us that there is a solid wooden table in front of us, physicists tell us there exists a lattice of microscopic particles, punctuated by comparatively large areas of empty space. Where commonsense tells us that rational agents act on their beliefs and desires, science tells us that brain events take place that cause their actions. Philosophical defenders of commonsense do not dismiss science in the same way that they dismiss anti-realist metaphysics. They typically argue for the compatibility of science and commonsense, insisting that the table of commonsense is, despite appearances, identical with the table of physics. But even a cursory knowledge of current science ought to make us aware that this compatibility is far from unproblematic. Physicists tell us that the fundamental particles of physics are not the small, solid, enduring, spatio-temporally located particles of commonsense, but are in fact a mixture of particles and waves vaguely distributed through a region. It is not obvious that this scientific conception of matter can be fully reconciled with the commonsense conception of matter.

Wilfrid Sellars is one philosopher who doubted that commonsense and science could be fully reconciled. He suggested that commonsense should give way to science in so far as the two prove to be incompatible. The argument is to be found in his highly influential essay, 'Philosophy and the Scientific Image of Man' (Sellars 1962). There he argues that we possess two unreconciled 'images' of man-in-the-world. These are the 'manifest image', and the 'scientific image'. Sellars deliberately chose the word 'image' because it is 'usefully ambiguous' (1962, p. 41). However, since he holds that the aim of philosophy is to '... understand how things in the broadest possible sense of the term hang together in the broadest possible sense of the term.' (1962, p. 37), it would seem that an understanding of the two images as world-views, or explanatory frameworks, cannot be too far from what he intends.

Sellars' manifest image is not commonsense itself, but the refinement of commonsense that philosophers and other intellectuals have developed over the centuries. Sellars regards most philosophy as being in the business of refining the commonsense view of the world. He nominates the work of analytic philosophers influenced by the later Wittgenstein as being the philosophy that has 'increasingly succeeded in isolating [the manifest image] in something like its pure form' (1962, p. 51). If the manifest image were to have to be modified or discarded due to the influence of science, then, because the manifest image is a refinement of commonsense, it appears that we would have to modify or discard commonsense along with it.

The scientific image is not science itself, but our current conception of the world that science makes present to us. The scientific image is an image of a world of microscopic particles and fields and forces, obeying statistical laws. The scientific image accounts for the manifest image, but only to an extent. We can provide a scientific explanation for why the sun appears red when it sets: The light rays, which we observe, are emitted by the sun at particular frequencies; these are frequencies that are then affected by the Earth's atmosphere in different ways, depending on their particular angles relative to the Earth's surface. By accepting this explanation we appear to sacrifice our ability to provide an account of the phenomenal character of redness, as it is experienced by us, in terms that are unified with the terms used in our explanation of the behaviour of light. Similarly, an explanation of how mental activity is caused by brain events is not in itself an explanation of our own experiences of conscious deliberation. It seems that some of the most important qualities of the manifest image have no counterparts in the scientific image.

We might, as scientific realists often hope, be able to integrate the manifest and the scientific images, or we might reject one in favour of the other. Currently, though, we do neither of these things. Instead we flit back and forth between the two unintegrated images. When we look at a sunset we admire the beauty of its rich red hues. When we explain the appearance of the sunset we are satisfied to say that, for most people, redness happens to covary with a complex of particular frequencies of light. The manifest image and the scientific image observe an uneasy truce, but it is not obvious that this truce can last in the long term. As the scientific image comes to assume an ever greater importance to us, it becomes ever more tempting to reject those aspects of the manifest image with which it cannot be reconciled. Eliminativist philosophers, such as the Churchlands (1998), urge us to give in to this temptation. If commonsense talk of belief and desires has no apparent place in the scientific image, so much the worse for talk of beliefs and desires, according to eliminativists. Better, they maintain, that we retrain ourselves to describe our inner life in the scientific language of brain science, than remain trapped by the archaic language of the manifest image.

In recent times the reputation of commonsense reasoning has been undermined, to an extent, by the investigations of psychologists. Reid assured us that the reliability of 'natural judgement' is self evident; however Tversky and Kahneman's examinations of lay judgments in statistics appear to demonstrate that, left to its own devices, natural judgement will lead to a disturbing variety of statistical errors. For example, natural judgement is guilty of continually overestimating the likelihood of conjunctive events and continually underestimating the likelihood of disjunctive events (Tversky and Kahneman, 1974). Social psychologists, such as Ross and Nisbett (1991) allege that our natural judgments about our fellows are systematically in error. We persistently suppose that the personalities of other people make a greater causal contribution to their behaviour than available evidence suggests. A third area in which natural judgement has been found wanting, by psychology, is in folk physics. Our natural judgement is that, if we impart a force on an object by tying it to a string and accelerating it in an arc, it will continue to move in an arc when we let go of the string (McCloskey 1983). It is no surprise that natural judgement does not accommodate the deep mysteries of quantum mechanics. However, it comes as something of a shock to discover that natural judgement has not even managed to accommodate itself to Newtonian mechanics, remaining partially in the grip of a naive impetus theory of motion.

Against those who charge that the natural judgment of commonsense is importantly defective, there are two main lines of defence. The first is to reject the coherence of the charge. Davidson (1984), and those influenced by him, argue that, in order to meaningfully interpret behaviour at all, we have no choice but to assume that others are substantially rational and that their beliefs are for-the-most-part true, as are our own. If Davidsonian views about interpretation are right then it seems that there is no possibility of consistently mounting an attack on the rationality of commonsense. We are trapped into believing that commonsense judgments are rational, on pain of undermining our own claims to be arguing rationally.[6]

The second line of defence, which has been pursued by Gigerenzer et. al. (1999), is to downplay the importance of the various experiments that purport to demonstrate significant failures of human reasoning. Our performance in solving some abstract problems has been shown to be poor, but the extent to which these failures are representative of human reasoning in situations that we actually encounter is an open question. It may be that our failures in abstract experimental situations are the minor side effects of heuristics that enable us to succeed in solving real-life problems.

Science began as an extension of commonsense, and it retains strong continuities with commonsense reasoning. Scientists are typically not trained to practice methods that are at odds with commonsense. However, science gives us a platform from which to challenge our commonsense judgements. Perhaps commonsense will turn out to be a ladder that we discard once we fully embrace the scientific image. As science develops, so does the scientific image, and so do tensions between it and the manifest image. So too do threats to the reputation of commonsense psychology and to the plausibility of commonsense realism.

In part, contemporary science is in conflict with commonsense. Because scientific realists seek to justify our belief in the claims of successful science, contemporary scientific realism also runs into conflict with commonsense. However, scientific realism is committed to the commonsensical view that there are mind-independent objects residing in an 'external' world; and the proponents of scientific realism appeal to commonsense practices to justify their use of abduction. So there is a tension in contemporary scientific realism. Scientific realism both conflicts with and relies on commonsense. It is becoming increasingly important for scientific realists to address this tension and clarify the extent to which they remain committed to the conceptual framework and inferential methods of commonsense.

The theme of the enduring place of commonsense reasoning in science, and indeed in debates about scientific realism, is exemplified in several of the papers in this collection. Harold Kincaid argues that the case for realism in given domains of scientific practice is one that is established contextually. This is because, he holds, scientific inference is empirical and social in ways that are not captured by the 'logic of science' ideal. If Kincaid is right that there is no universal logic of science, then it would appear that scientific inference cannot, in general, be sharply distinguished from commonsense reasoning.

Robert Nola defends the commonsensical appeal to manipulability as a mark of the real, as does Steve Clarke (2001). This appeal is at the heart of most cases for entity realism. Entity realism is commonly associated with Ian Hacking's (1983, p. 23) catchcry, 'if you can spray them then they are real', however its roots lie in the anti-theoretical traditions of pragmatism and in the commonsensical reaction to Berkeley's idealism made famous by Samuel Johnson. When Johnson manipulated a stone with his foot he at once affirmed that the stone was more than a mere theoretical construct and that he was warranted in applying commonsense reasoning to affirm its existence.

We have already seen the tensions between science and commonsense expressed in Sellar's talk of the scientific and the manifest image. In the tradition of Peirce and Dewey, Herman De Regt takes a pragmatist point of view, regarding talk of the two images. He argues that the two images are reconciled in a scientific attitude properly understood as being grounded in commonsense. Brian Ellis seeks to recast talk of the two images in the terminology of the new essentialism, which he advocates (Ellis 2001). Ellis argues that the aspects of the manifest image, such as consciousness, that have traditionally been thought to be lost in the scientific image can be understood as part of the scientific image after all, when we adopt essentialism.

A general problem for defenders of commonsense is presented by the psychological evidence already discussed, which threatens to undermine the credibility of our everyday reasoning. A specific problem for the characterisation of a defensible version of commonsense psychology is presented by the phenomenon of delusion. Lisa Bortolotti argues that delusions are best understood as irrational beliefs involving resistance to contrary evidence and compartmentalisation. Her argument serves to narrow the gap between the epistemology of delusional and non-delusional believers. In doing so it contributes to the analysis of commonsense psychology.[7]

Indiana University — Purdue University Indianapolis (IUPUI)

and

Centre for Applied Philosophy and Public Ethics, Charles Sturt University

REFERENCES

Boyd, R.: 1973, 'Realism, Underdetermination, and a Causal Theory of Evidence,' *Nous* 7, 1-12.

Carrier, M.: 1991, 'What is wrong with the Miracle Argument?' *Stud. History and Philosophy of Science,* 22, 23-36.

Carrier, M.: 1993, 'What is Right with the Miracle Argument: Establishing a Taxonomy of Natural Kinds,' *Stud. History and Philosophy of Science.* 24, 3, 391-409.

Churchland, P.S. and Churchland P.M.: 1998, *On The Contrary: Critical Essays 1987-1997*, The MIT Press, Cambridge MA.

Clarke, S.: 1997, 'When to Believe in Miracles', *American Philosophical Quarterly* 34, 95-102.

Clarke, S.: 2001, 'Defensible Territory for Entity Realism', *British Journal for the Philosophy of Science* 52, 701-722.

Davidson, D.: 1984, *Inquiries Into Truth and Interpretation*, Oxford University Press, Oxford.
Duhem, P.: 1954 [1906], *The Aim and Structure of Physical Theory*. Trans. Philip P. Wiener, Princeton University Press, Princeton.
Ellis, B.: 1979, *Rational Belief Systems*, Blackwell, Oxford.
Ellis, B.: 1985, 'What Science Aims to Do,' in, *Philosophy of Science: Images of Science*, edited by P. Churchland and C. Hooker, University of Chicago Press, Chicago.
Ellis, B.: 1990, *Truth and Objectivity*, Blackwell Cambridge.
Ellis, B.: 2001, Scientific Essentialism, Cambridge University Press, Cambridge.
Fine, A.: 1984, *The Shaky Game*, University of Chicago Press, Chicago.
Forguson, L.: 1989, *Commonsense*, Routledge, London.
Gigerenzer, G., Todd, P.M. and the ABC Research Group.: 1999, *Simple Heuristics That Make Us Smart*, Oxford University Press, Oxford.
Kitcher, P.: 1993, *The Advancement of Science*, Oxford University Press, Oxford.
Kukla, A.: 1998, *Studies in Scientific Realism*. Oxford University Press, Oxford.
Laudan, L.: 1977, *Progress and its Problems*, Oxford University Press, Oxford.
Laudan, L.: 1981, 'A Confutation of Convergent Realism,' *Philosophy of Science* 48, 19-49. Reprinted in *Philosophy of Science*, edited by D. Papineau (1996), Oxford University Press, Oxford.
Laudan, L.: 1985, 'Explaining the Success of Science,' in J.T Cushing et al. (eds.), *Science and Reality*, University Notre Dame Press, South Bend.
Lipton, P.: 1993, *Inference to the Best Explanation*, Routledge, London.
Lipton, P.: 1994, 'Truth, Existence, and The Best Explanation,' in *The Scientific Realism of Rom Harré*, edited by Anthony A Derkson, Tilburg University Press, Tilburg.
McCloskey, M.: 1983, 'Intuitive Physics', *Scientific American* 248, 122-130.
Musgrave, A.: 1985, 'Realism versus Constructive Empiricism,' in *Philosophy of Science: Images of Science*, edited by P. Churchland and C. Hooker, University of Chicago Press, Chicago.
Musgrave, A.: 1988, 'The Ultimate Argument,' in *Relativism and Realism in Science*, edited by Robert Nola, Kluwer, Dordrecht, pp. 229-252.
Newton-Smith, W.: 1981, *The Rationality of Science*, Routledge, London.
Peirce, C. S.: 1958, *Collected Papers*. Vol. 5, Harvard University Press, Cambridge.
Psillos, S.: 1999, *Scientific Realism: How Science Tracks Truth*, Routledge, London.
Putnam, H.: 1975, *Mathematics, Matter and Method*, Cambridge University Press, Cambridge.
Putnam, H.: 1981, *Reason Truth and History*, Cambridge University Press, Cambridge.
Putnam, H.: 1982a, 'Why There isn't a Ready Made World,' *Synthese* 51, 141-67
Putnam, H.: 1982b, 'Why Reason can't be Naturalized', *Synthese* 52, 3-23.
Putnam, H.: 1984, 'What is Realism?', in *Scientific Realism*, edited by Jarrett Leplin, University of California Press, Berkeley.
Quine W.V.O.: 1975, 'On Empirically Equivalent Systems of the World' *Erkenntnis* 9, 313-28.
Rescher, N.: 1987, *Scientific Realism: A Critical Reappraisal*. D. Reidel, Dordrecht.
Ross, L. and Nisbett, R.: 1991, *The Person and the Situation: Perspectives of Social Psychology*, McGraw-Hill, New York.
Sankey, C.H.: 2001, 'Scientific Realism: An Elaboration And A Defence', *Theoria* 98, 35-54.
Sellars, W.S.: 1962, 'Philosophy and the Scientific Image of Man', in R.G. Colodny (Ed.), *Frontiers of Science and Philosophy*, The University of Pittsburg Press, Pittsburg, pp. 37-78.
Sellars, W.S.: 1962, *Science, Perception, and Reality*, Humanities Press, New York.
Smart, J.J.C.: 1968, *Between Science and Philosophy*, Random House, New York.

Stein, E.: 1997: 'Can We Be Justified in Believing That Humans Are Irrational?', *Philosophy and Phenomenological Research* 57, 545-565.
Tversky, A. and Kahneman, D.: 1974, 'Judgement Under Uncertainty: Heuristics and Biasses', *Science* 185, 1124-1131.
Whewell. W.: 1840, *The Philosophy of the Inductive Sciences*, J.W. Parker, London.
Worrall, J.: 1989, 'Structural Realism: The Best of Both Worlds?' *Dialectica* 43, 99-124.
Van Fraassen, B.: 1980, *The Scientific Image,* Oxford University Press, Oxford.

NOTES

[1]Although Putnam (1981) (1982a) (1982b) and Ellis (1990) (1996) advocate an epistemic notion of truth, they both consider themselves to be scientific realists.

[2]We use the term 'most' here to allow for exceptions such as 'perfectly reversible heat engines.'

[3]For an argument to the conclusion that appeals to the miraculous can be explanatory see Clarke (1997).

[4]In fact, non-realists, such as Laudan (1981), point out that if we treat the argument as a deduction it commits the fallacy of affirming the consequent: if (P), then (Q); (Q); therefore, (P). This can be seen most easily by focusing on 1 and 2 and reversing their order.

> 2: If our theories were (approximately) true, (P), then their success, (Q), would be a matter of course.
>
> 1: Our theories are successful, (Q)
>
> ---
>
> Therefore, (probably) our theories are (approximately) true, (P).

Some say the argument can be made deductively valid (e.g., Musgrave, 1988), but this comes only at the cost of adding further contentious assumptions, e.g., a premise which states that we are justifed in believing that the best explanation of a phenomenon is true.

[5] Here we are following Forguson, (1989, Chapter 1).

[6]The charge that attempts to identify widespread human irrationality must be self-undermining is further explored by Stein (1997).

[7]Thanks to Keith Horton for useful comments on draft material.

ROBERT NOLA

REALISM THROUGH MANIPULATION, AND BY HYPOTHESIS

1. INTRODUCTION

Metaphysical, or ontological, realism with respect to the unobservables of science is the view that unobservable objects, properties of objects, events and processes exist independently of what we believe, perceive or say there is in the world, or what linguistic framework we use to talk about the world. Thus there are electrons, and they do have charge, spin and mass, they are involved in emission events during neutron decay, and enough of them moving in the right way constitute the workings of computers. Under this heading can be found the scientific entity realism advocated by Ellis, Devitt, Hacking, Cartwright and Hellman amongst many others.[1] Arguments will be canvassed in this paper in support of this kind of robust scientific realism. Such a realism stands opposed to varieties of anti-realism ranging from constructive empiricism (which remains sceptical, not only about the truth or verisimilitude of our theories even though the theories do have truth values, but also claims about what exists), to the constructivism of many sociologists of science.

This paper develops an argument for ontological realism with respect to those unobservable items that we manipulate in experimental and technological processes. This is called 'realism by manipulation'. Often items are manipulated in experimental setups in order to detect other items that are not manipulated. This will be called 'hypothetical realism' with respect to what is allegedly detected through the manipulation of something else. In arguments for both kinds of realism appeal will be made to hypotheses or theories about the manipulated and the detected. But there is a difference. In the case of realism by manipulation often only a single causal law or a claim about the causal power(s) of the manipulated need be invoked. In contrast any argument for the detection of items not manipulated may require much theory which is highly speculative or untested. Thus hypothetical realism about unobservables may often be less

S. Clarke and T.D. Lyons (eds.), Recent Themes in the Philosophy of Science, 1–23.

secure than realism based on manipulation because of the 'hypothetical' character of the theory about what is allegedly detected. But this is merely an epistemic matter that does not affect the ontological realism of either the manipulated items or the items merely detected. On the whole, support for a claim about the existence of some item based on manipulation is often stronger than support based on evidential considerations for hypothesised items. So, even though two kinds of realism are spoken about in the course of the paper, the division is between two kinds of evidence and argument for realism. This distinction is epistemic and is logically independent of ontological realism.

2. EXPERIMENTAL AND TECHNOLOGICAL PRAXIS AND A ROBUST REALISM

Most practicing scientists are involved with the experimental manipulation and the use, in technological applications, of not only observable items but also unobservable items postulated in scientific theories. Hacking is an advocate of realism with respect to unobservable entities, saying of positrons and electrons 'if you can spray them then they are real' (Hacking 1983, p. 23). In his 1983 book (pp. 22-4 and Chapter 16) Hacking describes two experiments in which an unobservable is manipulated to detect another alleged unobservable. In the first, electrons and positrons are sprayed onto a niobium ball suspended in a gravitational field at a temperature close to absolute zero. The point of the experiment is to detect the presence of hypothesised fractions of the charge e of electrons, such as $\pm 1/3e$, which would also indicate the presence of a free quark. There is some dispute amongst physicists as to whether the experiment does provide sufficient evidence for free quarks with fractional charges, as opposed to something else. Let us simply suppose that there is some evidence (Hacking cites some purported evidence for their existence (*ibid.*, p. 23). The point to be made is that the manipulated entities are the electrons and positrons; and in constructing the experimental apparatus appeal is made to just one of their simple causal properties, that of being able to neutralise any opposite charge. What could serve as a better mark of the real than such manipulability?

In contrast neither the fractional charges nor the quarks are manipulated; so no such strong mark of the real can apply to them. If there is some argument for their existence then it will be 'hypothetical' in that it appeals to, amongst other things, the speculative theory of fractionally charged quarks and evidence based on the frequency of their alleged detection (available theories of quarks suggest that the chance of their free occurrence will be very low, and so their detection involves dealing with very low probabilities). Such a hypothetical realism with respect to quarks and fractional charges is grounded, not only in experimental practice, but also in theoretical considerations and special modes of inference from which the existence of quarks can be inferred. Given a set of rival theories about the subatomic, including our best current quark theory, and the evidence gathered in detection experiments, inference to the best explanation might tell us that our best current quark theory, with its presupposition that quarks exist, provides the best explanation of what is observed. However such argumentation is of a quite different character from considerations based on experimental manipulation.

Hacking provides a second example — the building and use of a polarising electron gun, which was given the acronym PEGGY II (see Hacking 1983, Chapter 16). The task was to investigate weak neutral currents which were hypothesised to be carried by neutral bosons (a particle detected only much later than the introduction of PEGGY II). The job of PEGGY II was to produce left-handed and right-handed polarised electrons and to investigate the relative frequency of their scattering. Hacking's argument for realism with respect to (polarised) electrons is based directly on manipulability. Given these two different experimental manipulations of electrons, the grounds for realism with respect to electrons becomes even more epistemically secure. In contrast there is at best only a hypothetical realism with respect to bosons and the weak neutral current postulated in theories of physics. The hypothetical or tentative realism towards, say, the neutral bosons investigated using PEGGY II, would become more robust if we were able, in turn, to manipulate them for various experimental purposes. In such a stepwise manner we build up our increasingly realist picture of the universe.

Hacking does not argue that we first build an experimental apparatus and then later become realists about electrons because the apparatus is a reliable producer of electrons. Rather because scientists are realists about electrons and the causal powers that are attributed to them, they build the apparatus they do (e.g., PEGGY II) and then use it to investigate other phenomena:

> We are completely convinced of the reality of electrons when we regularly set out to build — and often enough succeed in building — new kinds of device that use various well-understood causal properties of electrons to interfere in other more hypothetical parts of nature (Hacking 1983, p. 265)

We need to investigate what reasons there are for being 'completely convinced' about the prior existence of electrons and their casual powers. And unlike Hacking's argument, it will be argued in Section 4 that the manipulation of electrons can be parlayed into an argument for their reality.

In the initial discovery of electrons by J. J. Thompson in the late 1890s, electrons were not manipulated but merely detected as they came off the cathode (i.e., negatively charged plate) in cathode-ray tubes. The cathode ray tubes were elongated, and at the end opposite the cathode there was a fluorescent screen that would brighten when electrons struck it. However it became possible to indirectly manipulate electrons. Thompson placed charged plates surrounding parts of the tubes to create an electric field in order to deflect the beam of 'cathode-rays' or electrons (as they were subsequently called). Depending on the amount of deflection, as measured by means of the fluorescent screen, he was able to determine the mass/charge ratio. Here what is directly manipulated includes the cathode from which electrons were discharged and the electric field between the charged plates. It is this field which interacts with the electrons as they pass along the tube. It was previously well known that electric fields existed and could deflect charged particles; their use as manipulators is a mark of realism with respect to such fields (whatever their nature be, something of which we might know little). The electrons passing through the field are affected by it, and so are deflected from their straight line path down the tube (by an angle that can be measured on the fluorescent screen). So indirectly, the electrons can also be manipulated by the electric field. But since they are not directly manipulated in order to detect some other item, at best the electrons have, in Thompson's apparatus, a hypothetical realism. Latter when they can in turn be directly manipulated their realism becomes more robust.

Manipulation occurs not only in the physical and chemical sciences but in most other sciences as well. Perhaps the most recent spectacular example of manipulation is in biochemistry, where genes, or more accurately sequences of DNA molecules, are manipulated for a host of ends in a growing gene technology. Here the goal of manipulation need not always be more experiment; in the case of gene technology the ends are increased productivity of crops or greater disease management. We can adapt Hacking's remark to say, in the case of genes: 'we are completely convinced of the reality of genes when we regularly set out to devise — and often enough succeed in devising — new kinds of techniques that use various well-understood causal properties of genes for a range of desired ends, from health to food production'. Such manipulability of the unobservable remains within the realm of pure science (or 'blue skies' research) when it occurs in furthering successful experimentation; but it reaches outside pure science when it enters into technologies that require reliable means for the production of items that are either individually or socially desired. Another technological example is the use of lasers in our CD players. Lasers are a product of laboratory experiments that can now be manipulated by investigators in all sorts of experimental investigations. And their manipulable features can be 'encapsulated' as when they are built into our ordinary homely CD players.

At best manipulability is a mark of the real and not a criterion for the real. Manipulation is a success term (one cannot manipulate the unreal), and is a sufficient condition for realism; but it is not a necessary condition. The free quarks or fractional charges mentioned in the above experiment are (at this time) not manipulable by us (rather than, say, merely detected by us). If we are to argue for their realistic status, then we will have to do that by other means — unless, of course, we find a means of manipulating them. However not everything that is real is manipulable by us, e.g., pulsars, galaxies and black holes. And if current quark theory is correct and quarks do exist, then they may never be manipulable by us. Nor can we manipulate tectonic plates even though they have devastating effects upon us as they grind against one another around the surface of the planet. The realism associated with manipulability arises from the realism with respect to the relata which stand in causal relations; but only in some cases can the causal relations be exploited by us through manipulation. In the next section this will be explored further.

Note that we need not have full knowledge of what we manipulate in order to bring about a desired end. Perhaps we have only knowledge of a few of the causal powers of the items manipulated to bring about some effect. This is the case with the electrons used in the fractional charge experiments and in PEGGY II; and it is also the case with the electric fields used in Thompson's early work on the electron. We do not know, yet, what is the full nature of electrons, or electric fields, or what might be all the laws which apply to them. Related points can also be illustrated in the case of Semmelweis' discovery of 'cadaveric particles'.

Semmelweis was able to reduce the incidence of death by childbed fever in the Vienna hospital in which he worked by getting the student doctors, not merely to wash their hands, but to wash them in a chloride of lime solution. He hypothesised that 'cadaveric particles', as he called them, remained on the hands of doctors who examined corpses before they examined women about to give birth, thereby infecting the women with childbed fever. Through the new hand-washing procedures Semmelweis was able to manipulate the frequency with which 'cadaveric particles' were present on the hands of doctors, largely reducing it to zero. In turn he was able to manipulate the frequency with which 'cadaveric particles' affected women, and thus manipulate the frequency with which death by childbed fever occurred. It was the spectacular lowering of the incidence of childbed fever that lead people to believe that Semmelweis had made an important discovery, not only in the prophylactic procedures of hand-washing, but in medicine itself, viz., the discovery of some kind of infectious particle.

Semmelweis' discovery occurred in the absence of much knowledge about disinfectants, in particular how chloride of lime bought about its effects. And it occurred without any well-developed germ theory of infection, or any knowledge of what the 'cadaveric particles' were like. In fact Semmelweis was even wrong about the claim that childbed fever came only from 'particles from dead bodies', and he realized this once he discovered that even infected living tissue could also cause childbed fever. In the light of this, he subsequently referred to 'decaying animal-organic matter'. But this extensive ignorance of the agent (or kinds of agent) causing childbed fever did not prevent him from manipulating its frequency thereby reducing the incidence of childbed fever. And here it is such ability to manipulate, even in the absence of knowledge about what was being manipulated, that is a strong mark of realism with respect to even the initially misnamed 'cadaveric particles'. We owe it to much later investigators to tell us about the kinds of bacteria (particularly *streptococci*) which caused childbed fever.[2]

The above example, as well as many other examples from science, show how we can have changing scientific beliefs about the same (kind of) thing (e.g., electrons, streptococci, etc.) while reference to that (kind of) thing remains invariant. At the time experimenters first come into contact with some item they might have entertained a mixture of true and false beliefs about what they had discovered. (Thus Semmelweis held false beliefs about his 'cadaveric particles', and Thompson briefly held the belief that electrons existed in matter as do cherries in a cherry cake.) But as time passes these false beliefs get discarded and new beliefs are acquired (which in turn may be either true or false). Such change in belief about the same items is to be sharply distinguished from change in belief about what items exist. In some cases scientists do change their beliefs about what exists; they no longer talk of phlogiston or caloric (as in theories of heat). But we must not entertain such an exaggerated picture of this when we view the scientific community as talking of 'two different kinds' of entity, such as the Thompson-electron and the Bohr-electron, or the 18th-century-Sun (Isaac Newton regarded it as a ball of fire) and the 20th-century-Sun (a ball of nuclear fusion). The best explanation for what is going on in these cases is not that scientists changed their beliefs about what exists; rather it is the very same item about which they earlier believed one thing and later believed another. As with theories of the sun, it is the same items, electrons, that both Thompson and Bohr talked about in their quite different theories. What we need but cannot give here,[3] is an account of the invariance of the reference of scientific terms with variance of theory about the items referred to.

If we allow that there can be changing theories about the same scientific entities, then this has important negative implications for the pessimistic meta-induction expressed by Putnam: 'just as no term used in science of more than fifty (or whatever) years ago referred, so it will turn out that no term used now (except maybe some observation terms) refers' (Putnam 1978, p. 25). As it stands this a highly exaggerated claim. Modifications to it, to make it more plausible over time, yield only a meagre harvest of terms as compared with the terms of science that have long continuity of reference. Moreover the claim does not spell out, and depends upon, theories about how the meaning and reference of terms can be fixed which avoid some of the less plausible aspects of Kuhnian and Feyerabendian incommensurability. Once it is acknowledged that the pessimistic meta-induction must be accompanied by a theory about the meaning and reference of scientific terms, and that we must allow for changes of belief about the same kind of entity with change in theory, then the pessimistic meta-induction over the past history of theories looses much of its punch.[4]

There is also an important lesson here for ontological realism and arguments for it. Realists must also allow that there can be changes of belief about the same entity with change in theory. So no argument for realism should aim to show, using inference to the best explanation or any other inference pattern, that our theories are strictly true. This shows not only much more than realists need, but often something that is implausible. More cautious realists aim to show that our theories have some verisimilitude, from which it is inferred that the entities postulated by the theory exist, thereby also establishing ontological realism. Just how well realists can replace talk of truth by talk of truthlikeness and still preserve all the inferences they would like to make is a matter extensively explored by Niiniluoto (see Niiniluoto 1999, section 6.4 and 6.5). There he also explores in particular just how well we can make an acceptable inference from the success of a scientific theory to its truthlikeness. Thus there are forms of inference to the best explanation of which realists should be wary, viz., those which conclude that a theory is strictly true. But there are other forms that draw different, weaker conclusions which realists can use and which are effective for their purposes.

3. CAUSATION AND MANIPULATION

Hacking's distinction between entity realism ('a good many theoretical entities exist') and realism about theories (which 'says that scientific theories are either true or false' (Hacking 1983, p. 27)) is of a piece with the account of realism adopted here. And like any good ontological realist he also envisages that reality is 'bigger' than us because there might exist many more entities than those we actually have come to know about, or are possible for us to come to know. Some subset we come to manipulate, and another subset we get to know about *via* detection but do not, and perhaps never can, manipulate.

We can also adopt the following characterisation of 'actuality' (for our purposes no distinction need be drawn between the actual and the real) given by Frege: 'The world of actuality is a world in which this acts on that and changes it and again undergoes reactions itself and is changed by them' (Frege 1977, p. 27). We can treat the actual (i.e., the real) world as a causal nexus in which each item can be causally acted upon by some other item. If our world has free quarks and fractional charges, just as it has electrons and positrons, then they will enter into a causal nexus with one another such that we can use the latter to detect the former. It might be objected that Frege's criterion ignores the possibility that this world contains inert items which fail to enter into any causal interaction with any other item (or inert systems of objects which while intra-acting with one another do not inter-act with our system). So reality is 'bigger' than us in that it has entities that, because of their inertness, we can neither detect nor even manipulate. Setting this possibility aside, what Frege's criterion requires is that all items (other than the inert) do at least some work in causally interacting with some other item, or allow themselves to be acted upon by at least one other item, thereby entering minimally into some causal interactions. And this is all we need for the possibility of our detecting any entity.

In the above, causal relations play a crucial role. What is it for one thing to causally act on another? Though there are many theories of causation, there is one due to Gasking (1996) which is particularly germane to the idea of manipulation as a mark for the real. Gasking resists the non-realist view that causation is to be understood either as a pattern of inference or merely as a generalisation from which inferences can be made. Instead he uses as his models for causation those in which we use manipulative techniques, such as our making an iron bar glow by heating it, or producing a fire by rubbing sticks near dry grass, etc. Of these cases Gasking says:

> When we have a general manipulative technique which results in a certain sort of event, A, we speak of producing A by this technique. (Heating things by putting them on fire.) When in certain cases application of the general technique for producing A also produce B, we speak of producing B by producing A. (Making iron glow by heating it.) And in such a case we speak of A causing B, but not vice versa. Thus the notion of causation is essentially connected with our manipulative techniques for producing results. ... We could come rather close to the meaning of 'A causes B' if we said: Events of the B sort can be produced by means of producing events of the A sort'. (Gasking 1996, pp. 110-1)

Later Gasking goes on to say that he has made two points. The first is that he has 'explained the "cause-effect" relation in terms of the "producing-by-means-of relation"'. The second point is that he has given 'a general account of the producing-by-means-of relation itself'. And the general account is based on the fact that 'we learn by experience that whenever in certain conditions we manipulate objects in a certain way a certain change, A, occurs' (*ibid*., p. 114).

Gasking has said variously that the cause-effect relation has been *explained* by the producing/manipulation relation, or that these relations come close in *meaning*, or that they are '*essentially* connected'. Not all of these can hold together. And in fact perhaps none of them strictly hold. Moreover there are many things that can stand in a cause-effect relation that do not stand in a 'produced-by' relation based in manipulation. Further, it might be hard to see how the produced/manipulation relation itself might be cashed out without appeal to causation.[5] But these critical points aside, Gasking makes a suggestive link between the model of ourselves as possessors of manipulative techniques for bringing about A, or producing some change, A, and the further idea of A being caused. And it is this close link between manipulation and causation which will be exploited in what follows. Our experimental techniques give us new means of human manipulation; and on the basis of Gasking's suggestion, we can appropriately think of this as a cause-effect relation. This has been illustrated in the above examples of experimentation. We bring about change A, viz., we spray electrons and positrons, or we produce a stream of polarised electrons. And the technique of producing A also produces B; that is, in appropriate conditions we successfully detect fractional charges, or we detect weak neutral currents.

4. THE ARGUMENT FOR ONTOLOGICAL REALISM FROM MANIPULABILITY

What argument is there from the manipulability of some entity E in an experimental setup to its existence? It might be thought to be fairly direct because of the success character of the word 'manipulate'. One can not manipulate the nonexistent; if one successfully manipulates E then E exists. But there are further considerations in the case of the successful manipulation of unobservables in experimentation, or in the successful use of unobservables in technologies. These are set out below.

Let us suppose of some entity E (such as electrons) the following four claims:

(1) E exists and has causal power(s) P.

For example, in the experiment concerning the detection of free quarks one need merely assume the existence of electrons and positrons, something about their mode of production, the fact that they are negatively (or positively) charged and the causal powers that arise from this, and how this will affect the overall charge on the niobium ball causing it to move in the electric field.

(2) We have theory T about postulated entity F (e.g., quarks with fractional charges).

(3) On the basis of hypothesised causal power P of E and theory T of entity F, we can build an apparatus which exploits E's power to P in various circumstances, leading to the detection of Fs via the manipulation of Es.

This we may assume poses both a technological problem for experimenters in building an apparatus and a problem in establishing a theory about the apparatus. In addition experimenters must also eliminate any 'bugs' and any interference from extraneous factors (usually by a trial-and-error process), thereby building a reliable detector of Fs.

(4) The apparatus does produce some success in detecting Fs (e.g., fractional charges).

We can now ask: what explains (4), the successful use of the apparatus to detect some Fs? Before answering this, consider the possibility of experimental failure. We have some idea of what would underpin the failure to detect fractional charges, or the failure to detect them in the anticipated amounts (finding a lot fewer, or far more, than were anticipated). The failure could be due to the faulty theory T about Fs. Either T is wrong about there being Fs in the first place; or it is right that there are Fs but it says some wrong things about them thereby leading to our inability to detect them at all or detect them in roughly the right amounts.[6]

Alternatively the failure to detect Fs might be due to the apparatus, its construction, the failure to eliminate its 'bugs', or the theory of its functioning. Failure can also arise in other ways. Call claims like (1) and (3) collectively 'the manipulability claims M'. We might not be manipulating in the apparatus the very electrons and positrons we thought we were manipulating. Or we might be wrong about claim (1) in the following ways: either we are wrong about the existence of the Es (e.g., electrons) we purport to be manipulating, or, if we are right about their existence, we are wrong about the causal power(s) we attribute to them. Or we could locate the fault in an aspect of claim (3), viz., the way in which the Es (e.g., electrons with their causal powers) are alleged to interact with the Fs (e.g., the hypothesised fractional charges).

In contrast suppose the experiment is successful in the way indicated in (4). What explains the success? In part this has to do with the correctness of the manipulability claims. We could envisage an abductive argument of the following sort advocated by Peirce (Peirce 1901, p. 151):

> The surprising fact, C, is observed [the success mentioned in (4)];
>
> But if A were true [e.g., manipulability claims M in (1) and (3)], C would be a matter of course;
>
> ---
>
> Hence, there is reason to suspect that A [manipulability claims M] is true.

Thus we can infer conclusion, A, concerning the manipulability claims, and in particular the existence of entity E (e.g., electrons) with the causal power(s) P we attributed to E for the purposes of the manipulation. We cannot infer anything else about electrons, or any theory about them. But is this abductive inference both sound and strong? Its soundness turns on the correctness of the second premise, a counterfactual which we may suppose is backed by either a causal law or the operation of some causal power. Also, to accept this claim, we need to show that there is not some other causal hypothesis B that rivals A, in that it would also bring about C. That is, we need to show that cause A, when compared with all its rival causes B, B', etc., is the most probable cause of C. That is, we need to show that the manipulability claims M contain the best causal recipe for bringing about C.

In the light of this, the abductive argument needs to be reformulated along the following lines:

> The surprising fact, C, is observed [the success mentioned in (4)];
>
> A [e.g., manipulability claims M in (1) and (3)], has a sufficiently high probability as the cause of C, i.e., p(C, A) is sufficiently high;
>
> For any other cause B, p(C, A) > p(C, B);
>
> ---
>
> Hence, (there is reason to suspect that) A [manipulability claims M] is the correct causal hypothesis.

(Note that the second premise is needed to ensure that A is not simply the best of a bad bunch of hypotheses and that it meets some threshold of sufficiently high probability). This adaptation of Peirce's argument-scheme has some features in common with inference to the best explanation (IBE). However the conclusion is not the bare claim that some theory is true; rather it is that some supposed cause is the correct cause (or as in Peirce's conclusion 'we have reason to suspect that C is the cause'). Instead of IBE we have an example of inference to the most probable cause (IPC). This second form of inference, IPC, may well be acceptable while IBE is not, a point argued for by Cartwright.[7] Clearly for the revised form of abduction with a conclusion about the existence of electrons with particular causal powers, establishing the second premise is crucial.

Hacking resists asking what explains the success (expressed in (4)), saying:

> The argument — it could be called the experimental argument for realism — is not that we infer the reality of electrons from our success. We do not make the instruments and then infer the reality of the electrons .... That gets the time-order wrong ... we design apparatus relying on a number of home truths about electrons ... (Hacking 1983, p. 265).

His point is that experimenters simply accept that electrons exist with certain causal powers — and then they built an apparatus to exploit these powers. However Hacking's correct point about time-order does not strictly preclude any 'experimental argument for realism' such as that of IPC. Experimenters might have built their apparatus on the *assumption* that their 'home truths' about electrons were correct, and not on the bare 'home truths' themselves. So it still remains an open question as to whether the assumptions on which the apparatus has been constructed (electrons and their powers) are correct, even in the case where it successfully detects other entities (such as fractional charges). There remains room for an experimental argument for realism based on manipulation, even where there might be quite independent grounds for thinking that there are electrons with particular powers. The experimental argument would then add to such independent evidence about electrons and their powers.

Granting that the experimental argument for realism has the form of IPC, or Peircean abduction, as above, the conclusion — 'there is reason to suspect that the manipulability claims M are true' — might appear to be rather weak. Peirce's text does contain the clause 'there is reason to suspect' which modifies a contained claim, thereby weakening it to one of reasonable belief. However other formulations of Peircean abduction omit the qualification and simply boldly assert as the conclusion, 'the manipulability claims M are true'. Whichever way we take the conclusion, we do have a strong non-deductive argument, in the form of IPC, which does explain why the success mentioned in (4) has come about. And this conclusion based on success is to be contrasted with the possibility of failure in the many ways suggested above. In fact we could rephrase our question and call for the best causal explanation which answers the contrastive question: 'What accounts for the success of the apparatus, as indicated in (4), rather than its failure?' And here we need reasons for success which rule out failure.

The possibility of failure suggests another way in which in which the contrastive question might be answered that is not obviously a form of Peircean abduction. This is the 'no-miracles argument' (NMA) as outlined by Putnam (Putnam 1978, pp. 18-9). What is to be explained is the success of science S, rather than any of its failures. What is to do the explaining? Putnam envisages two hypotheses. Consider some theory T which is understood realistically in that it makes existential claims about the basic items it postulates (e.g., there are genes, curved space-time, etc.). Call the theory so understood 'T_R'. In contrast consider T interpreted 'positivistically' in which it is understood as a calculus, and in which the realistic assumptions of T_R are denied.[8] Call the theory so understood 'T_P'. Now both T_R and T_P entail the predictive and other successes of science S. Of these two interpretations of T, Putnam says: 'If there are such things [as indicated by T_R], then a natural explanation of the success of these theories is that they are partially true accounts of how they behave'. In contrast: '. . . if these objects don't really exist at all [as indicated by T_P], then it is a *miracle* that a theory which speaks of gravitational action at a distance successfully predicts phenomena' (*ibid*., p. 19).

Adapting NMA to the above considerations about manipulation claims M, we have the following two theories: M_R in which the manipulability claims are taken realistically, i.e., there are electrons with the supposed causal powers; and M_P in which the manipulability claims are taken non-realistically, i.e., there are no electrons and no such causal powers. Then on NMA, M_P makes the successful use of the experimental apparatus a miracle; none of the manipulation claims about electrons and their powers are true, yet the experiment does successfully detect fractional charges. In contrast M_R does explain the success of the experimental detection; the manipulation claims are true, and this is said to explain the success of the experimental detection. Here the kind of explanation is not mere deducibility from laws; it involves much more robust causal explanation in which there is a causal story about the production of electrons, their manipulation in the experiment and their detection of fractional charges, once the electrons (or positrons) have done their causal work. The conclusion of NMA is, 'there are electrons with particular causal powers'. Despite the strong appeal of NMA, some have found it a suspect form of reasoning. This will emerge again in Section 6; for those who find it suspect, a way of by-passing it will be suggested.

Hacking is right about the realist stance of scientists in building their apparatus in the way they do, by exploiting the causal powers of, say, electrons or positrons. But both IPC and NMA also indicate a way in which the realist stance of the experimenters is vindicated by a successful experimental outcome. Before their manipulation, Hacking appears only willing to give entities, such as electrons, hypothetical existence. Once we can manipulate them, their realism becomes much more robust. In order for this to happen, there needs to be some transition from a seemingly weaker realism, which is merely hypothetical, to a more robust realism based on manipulation. If there is no such transition then, for Hacking, the hypothetical realism with respect to these entities becomes suspect, and a certain scepticism towards them sets in. The particular case he mentions (somewhat prematurely) is that of black holes. He says in general: 'Long-lived theoretical entities, which don't end up being manipulated, commonly turn out to have been wonderful mistakes' (Hacking 1989, p. 275).

But need we be so agnostic? Can't we mount arguments for the existence of things like black holes, or neutrinos, which we cannot manipulate? We do admit of such things as pulsars and tectonic plates which we cannot manipulate; so the arguments for such things cannot turn on premises that involve appeals to certain entities with causal powers that are manipulated for particular ends. In the absence of IPC some have fallen back on various versions of IBE to argue the case for hypothetical realism about the non-manipulated entities detected in experiments. But not every argument for the existence of some entity need take the form of IBE. We need to consider those cases where evidence has a general bearing on existence claims (supposing, of course, that the claims of sociologists of 'knowledge' are not correct and that what determines what exists is entirely a matter of negotiation amongst scientists).

5. A PROBABILISITIC ARGUMENT FOR THE EXISTENCE OF NON-MANIPULATED ITEMS BASED ON EVIDENCE

This section considers arguments for realism with respect to items which are not manipulated but are at least detected (quarks, fractional charges, neutral weak currents, etc.), and for which there can be some evidence. Some of these items were postulated by theory and were not detectable at the time of their postulation — and some even now remain undetected (assuming they exist). Such was the case of Pauli's theory-based postulation of neutrinos in the early 1930's, which were not detected until the 1950s; or Schwarzchild's postulation of black holes shortly after Einstein's publication of his General Theory of Relativity, over which there still remains controversy as to whether they have been detected or not.

In some cases realism with respect to the postulated entities arises from the evidence gathered on behalf of the theory that entails the existence of such entities. There is a useful discussion of this in Dorling (1992) which takes a Bayesian approach to how evidence can bear on theories and the acceptability of either their realist or non-realist interpretation. In considering any theory T, Dorling invites us to distinguish the following: the realist interpretation of T, viz., T_R; its realist ontological commitments, R; the non-realist denial of R, viz., P (i.e., the 'positivist' commitment to not-R which is equivalent to P); and finally T's observational consequence class, T_P. T_P is an infinite conjunction of observational claims. It can also be regarded as expressing the empirical adequacy of T and as such it goes well beyond the current evidence for T in that it includes all of T's observational consequences, past, present and future. Thus incoming evidence E can give support to T_P in the same way it would to T_R. For our purposes we can regard T_P as the non-realist or 'positivist'[9] interpretation of T. Important for Dorling's purposes is the distinction between P, the denial of the realists' ontological commitments, and T_P which both the realist advocate of T_R and the non-realist are agreed upon. (Note that both the realist and the non-realist are agreed that T_R entails T_P. And note also that P does not entail T_P or its negation; so they are logically independent of one another.) Since Dorling is a subjectivist we are, in what follows, to read the probability expression 'p(X)' as 'the rational subjective degree of belief in X'.

Let us consider what happens as evidence E comes in. Suppose that the realist makes successful novel predictions (for example, the theory from which Pauli predicted the existence of neutrinos also makes other surprising testable predictions which turn out to be true). Then scientists are more willing to accept the ontological claims made by the theory; that is, as $p(T_P)$ rises on the basis of E, p(P) falls, which is equivalent to the claim that p(R) rises. Thus in situations of increasing positive evidence scientists tend to disbelieve the non-realist interpretation of their theory and start believing the realist interpretation. But suppose the theory T runs into trouble as E comes in. Then Dorling says '$p(T_P)$ falls and p(P) rises' (Dorling 1992, p. 364). However, matters are not wholly straightforward in this case. When T gets into trouble the blame may be sheeted home not to the ontological assumptions of the theory, but to what the theory says about what it postulates. That is, we need to take into account the matters raised at the end of Section 2 in which the entities we talk about in our theories can remain invariant while what we say about them can vary. So the alleged inverse relation between $p(T_P)$ and p(P)with adverse evidence E may not hold, or not hold in a straightforward manner. This aside, the significance of Dorling's point turns on our recognising the distinction between P and T_P and their independent variation.

Of related interest is Dorling's worked example (*ibid.*, pp. 368-9) of how a realist and a non-realist might assign subjective probabilities to their respective T_R and T_P (indicated by $p_{realist}(T_R)$, $p_{positivist}(T_R)$, and so on) and then revise their degrees of belief in the light of increasing positive evidence. Dorling's chosen numerical values show how, with incoming positive evidence the realist's degree of belief in T_R (and also R) will be strengthened. Meanwhile the positivist, who initially assigned a low subjective probability to T_R comes to assign a probability well above a half, and is thereby converted to the realist interpretation. In Dorling's worked example a positivist assigns a low subjective degree of probability of 0.2 to $p_{positivist}(T_R)$. But with increasing positive evidence E this becomes greater than a half, with the consequence that the scientist abandons the positivist interpretation of T in favour of a realist interpretation. However, if too low a probability is assigned by the positivist to $p_{positivist}(T_R)$ then the conversion would not occur.

This does represent how matters have occurred in the history of science, especially in the case of Daltonian atoms in chemistry, one of the cases Dorling uses to illustrate his worked examples. What is clear from this example is that support for realism can arise in the absence of any argument for realism based on manipulability; it turns on increasing positive evidence only. The opposite case of increasing negative evidence is not a straightforward matter and cannot be fully represented in the Bayesian terms given above. We also need to take into account accompanying semantic doctrines in which, even though we have negative evidence for T_R, the blame is sheeted home not to what the theory is about (its ontology) but what the theory says (its 'ideology').

Dorling's considerations are set within the framework of subjective Bayesianism. So it is open to non-realists who are subjective Bayesians to set their priors in any way they like. Subjectivists might assign such low values to realism and such high values to non-realism that the above case of conversion from non-realism to realism does not occur. However a less radically subjectivist account of probability such as Shimony's 'tempered personalism' might well allow a reasonable assessment of 'seriously proposed hypotheses' in which conversion either way is not a remote possibility (see Shimony 1993, Chapter 9, Section III). The kind of Bayesians envisaged here have a liberal point of view in that they are open to alternatives and can change their minds concerning realism *versus* non-realism on the basis of evidential information. Note that considerations of how evidence might bear on realist *versus* non-realist interpretations could be set in a non-Bayesian framework. However Dorling's Bayesian considerations do show how we can reach realist, or non-realist, conclusions about our theories on evidential grounds only.

6. SOME ARGUMENTS FOR THE EXISTENCE OF ENTITIES THAT ARE NOT MANIPULATED

In Section 4 we looked at IPC and NMA as forms of argument for realism with respect to manipulated entities. These argument forms have also been used to argue for a hypothetical realism concerning entities that are not manipulated. Also more full-blooded versions of IBE have been used to this end, as well as other forms of argument such as Norton's 'demonstrative induction' in the case of Bohr's arguments for the existence of electrons (see Norton 2000). However IBE has recently been criticised as an unsatisfactory form of inference. Its advocates regard it as a *sui generis* principle of non-deductive inference; its opponents claim that either it adds nothing to what can be gained from other probabilistic considerations, or, if it does add more, what it adds is either erroneous or incoherent.[10] Here we will focus on one limitation on the use of a form of IBE and contrast this with the kinds of consideration already mooted in Section 5 due to Dorling. In this final section two matters will be addressed: how some explanationist concerns are addressed in Bayesianism; and how the probabilistic considerations of Section 5 bear further on realism.

Consider Bayes' Theorem in the following form, where, as before, H is some hypothesis and E some evidence and 'p' stands for 'probability':

$$\text{(Bayes)} \qquad p(H, E) = p(E, H) \times p(H)/p(E) \quad (\text{where } p(E) \neq 0)$$

The factor 'p(E, H)' is called 'the likelihood'. This can be understood to stand for the relation of explanation where 'H' is the hypothesis doing the explaining and 'E' is what is explained. In the deductive Hempelian model this factor becomes 1; but in the Hempelian statistical model it will be less than 1. However understood as explanation, it is a rather indiscriminate factor in that it admits deductive, and thus explanatory, relations of the sort: X&E explains E. Even when the arbitrary X is a law this cannot count as an explanation of E. So even though p(E, X&E) = 1 it cannot be said that X&E explains E. Perhaps this is one reason why some explanationists would not want to adopt a Bayesian framework for their account of IBE. But let us set this aside and suppose that we rule out cases of this sort on the basis of other conditions on explanation.

Let us suppose we have two rival hypotheses H and K which do meet the conditions for being an explanation of E. Then can we infer anything about the *truth* of H or K? Or if we lower our sights, can we infer anything about the comparative degree of *support* that E gives to H and K? The former is unavailing, but the latter has some promise. So let us consider the following in which H and K do meet the minimal conditions for being satisfactory explainers. Then where we can say, of hypotheses H and K, that H is a better explainer of E than K, we will have at least the following:

(1) $p(E, H) > p(E, K)$. [by supposition]

Rearranging (Bayes) we have:

(2) $p(E, H) = p(H, E) \times p(E)/p(H)$

By appropriate substitutions-instances of (2) for H and K in (1) we have:

(3) $p(H, E) \times p(E)/p(H) > p(K, E) \times p(E)/p(K)$.

By canceling out the 'p(E)' factor (which is non-zero), we have:

(4) $p(H, E)/p(H) > p(K, E)/p(K)$; or

(5) $p(H, E)/p(K, E) > p(H)/p(K)$;

Now if the greater explanatoriness of H over K, as supposed in (1), is to carry over into greater confirmation of H by E than K by E, viz., the factor on the left-hand side of (5), then we have

(6) some factor greater than $1 > p(H)/p(K)$

from which it follows that

(7) $p(K) > p(H)$.

This argument tells us nothing about one of the common forms of IBE; from the assumption that H is a better explainer of E than K we can infer nothing about the truth of H. But if we lower our sights, can we infer that H is better supported by E than K? The above proof shows that this is so only on the condition that the prior probability of K is greater than the prior probability of H. That is, H explains E better than K, and so H gets better confirmation on E than K, only if H is initially less probable than K.[11] Is this a plausible requirement? As Rescher comments on his proof: 'the probabilistically better explanation is thereby the comparatively more acceptable only if it is also a priori *less* likely. But this clearly goes against the grain, seeing that it is implausible to make such a demand for explanatory shots in the dark' (Rescher 2000, p. 10). So IBE, understood not as an inference to the truth of some hypothesis but as an inference to the greatest probability on evidence of one out of a bunch of hypotheses, will only be acceptable on a condition which we need not always accept.

If probability theory does not fully vindicate the explanationist's role for explanatoriness as a premise in an inference to truth or confirmation, then perhaps probability can still be marshaled in support of realism. This has already been argued in Section 5 in which it was shown that a non-realist could, with incoming positive evidence, convert to realism. Initially it might appear that Bayesian considerations do not give support to realism since Bayesianism cannot support one version of the 'no miracles' argument (NMA). It is argued in Chapter 3 of Howson (2000) that NMA is in fact an unacceptable argument form open to counterexample. So NMA is unsatisfactory and cannot be used, as it was at the end of Section 4, to support even realism based on manipulation (but IPC can still do better). However in Chapter 8 in a section entitled 'A Sounder Argument for Realism' (*ibid.*, pp. 198-201), Howson uses probabilistic considerations raised by Dorling in support of realism (discussed in the previous section). The following is an expansion of these considerations which shows that there is no conclusive argument against a realist construal of our theories; it remains quite possible that $p(R) > p(P)$ where, as before, 'R' is the ontology of T understood realistically as T_R, and 'P' is the positivist denial of R (i.e., P is equivalent to $\neg R$).

Using the simple form of Bayes' Theorem (Bayes), as above, we have

(1) $p(T_R, E) = p(E, T_R) \times p(T_R)/p(E)$ (where $p(E) \neq 0$);

(2) $p(T_P, E) = p(E, T_P) \times p(T_P)/p(E)$ (where $p(E) \neq 0$).

Let us suppose that the same evidence E is entailed by both T_R and T_P. Then we have

(3) $p(T_R, E) = p(T_R)/p(E)$, and

(4) $p(T_P, E) = p(T_P)/p(E)$.

From (3) and (4) we can form the ratio

(5) $p(T_R, E)/ p(T_P, E) = p(T_R)/ p(T_P)$

Consider the right-hand side of (5). This is a ratio of prior probabilities which, we can suppose, remains fixed as a constant once the priors have been chosen. Writing the constant as k we now have:

(6) $p(T_R, E)/ p(T_P, E) = k$.

What (6) now shows is that with increasing evidence E the ratio on the left-hand side must remain equal to a constant. So as $p(T_P, E)$ goes up so must $p(T_R, E)$ in the same ratio to maintain the equality to the constant. Howson correctly comments: 'Whether T_R ever become more probable than not *a posteriori* will, of course, depend on its prior probability' (*ibid.*, p. 200). In this context Howson envisages a quite tempered view of how priors are to be assigned, one that precludes a rabid commitment to realism or positivism, ie., unreasonably low or high priors. That is, 'a non-zero prior is intended to model a state of mind that is prepared, even minimally, to be changed with the reception of suitable information, and we know … that a zero probability will not model such a mind' (*loc. cit.*). And this is surely the case for the liberally minded person who wants to know whether they can appropriately take their particular scientific theory realistically or not.

However more needs to be said about the above. The following is an important consideration based on the fact that the empirical content of T, T_P, is not something that the realist can set aside since T_R entails it: Since T_P is a logical consequence of T_R then by a theorem of the probability calculus,

(7) $p(T_R) < p(T_P)$

Using result (7) in (5) we have:

(8) $p(T_R, E) < p(T_P, E)$

This looks like bad news for the realist in that the probability of the realist construal of T must always lag behind its positivistic construal, as evidence comes in. But this is the wrong way of understanding what (8) says.

What (8) reflects is in fact the more risky content of the realist construal of T as compared with T's positivist construal, even in the light of evidence E. What we instead should consider is the fate of R and its positivist denial P; and (8) says nothing of this. In order to follow the fate of R and P, we need to follow, once more, Dorling's procedure of distinguishing carefully between R and T_R, as is commonly done, and P and T_P, as is not so commonly done. In fact P entails neither T_P nor $\neg T_P$. So from neither (7) nor (8) can one draw either of the following conclusions: (a) a pessimistic conclusion about R, the realist ontology of T_R (but note that T_R entails R); or (b) an optimistic conclusion, not about T_P (as might be done if the difference between T_P and P is ignored), but the anti-realist denial of R, viz., P (and note that P is logically independent of T_P). In fact what Dorling's numbers assigned in his worked example shows (outlined in Section 5) is that even when the positivist assigns a low probability to realist assumptions R, viz., $p_{positivist}(R) = .2$, and thus $p_{positivist}(P) = .8$, then as evidence E comes in the positivist can undergo a conversion to realism. And this becomes evident in the light of the above only when we track separately from the fate of T_P the quite different fate of P, the positivist denial of the realists' assumptions.

To see this note that T_P is the full true observational content of T. And as E comes in the probability of T_P can go up as it gets support from E; that is,

$p(T_P, E) > p(T_P)$. Also as E comes in we have $p(T_R, E) > p(T_R)$. But because of (6) each of $p(T_P, E)$ and $p(T_R, E)$ must go up in a constant ratio. However, nothing follows from this about the fate of P or $p_{positivist}(P)$. In fact $p_{positivist}(P)$ can go down in such a situation; and this is illustrated by Dorling's worked example (as we have seen in Section 5) in which $p_{positivist}(P)$ goes down so far with increasing E that conversion to realism occurs. Thus while (8) appears to be fatal for the realist it is not. Rather what one needs to track is the fate of $p_{positivist}(P)$ as E comes in; and (8) says nothing of this. Thus on probabilistic reasoning alone, and taking into account increasing evidential support, a case can be made for adopting realism with respect to the ontology of some given theory T. And a case can also be made for abandoning a non-realist understanding of T, if that was one's initial view of T.

The upshot of this section is that explanationists who use IBE as an independent principle might be able to provide grounds for the support of realism, using some appropriate form of IBE (that has not been criticised here[12]). However there are probabilists who deny that IBE has a worthy status and reject it, or at least place it under a heavy pall of criticism. But this does not preclude them from having other grounds for realism, as Dorling shows. So either with or without IBE, realism can still stand, providing realists and their opponents do not adopt a stance of uncompromising opposition to one another. And realists can do this, either for a robust realism based on manipulability, or a less robust hypothetical realism based on evidential considerations that do not have to do with manipulability. And such considerations help to undermine most non-realisms in respect of science, especially the position of (social) constructivists who can see no grounds for realism and adopt a constructivist account of the entities of science based on the consensus of a scientific community.

University of Auckland

REFERENCES

Ben-Menahem, Y.: 1990, 'The Inference to the Best Explanation', *Synthese* 33, 319-344.
Cartwright, N.: 1983, *How the Laws of Physics Lie*, Clarendon, Oxford.
Clarke, S.: 'Defensible Territory for Entity Realism', *The British Journal for the Philosophy of Science*, 52, 701-722.
Day, T. and Kincaid, H.: 1994, 'Putting Inference to the Best Explanation in Its Place', *Synthese* 98, 271-295.
Devitt, M.: 1997, *Realism and Truth*, Princeton University Press, Princeton, second edition.

Dorling, J.: 1992, 'Bayesian Conditionalization Resolves Positivist/Realist Disputes', *The Journal of Philosophy* 89, 362-82.
Ellis, B.: 1979, *Rational Belief Systems*, Blackwell, Oxford.
Franklin, A.: 2001, *Are there Really Neutrinos? An Evidential History*, Perseus Books, Cambridge MA.
Frege, G.: 1977, 'The Thought', in P. Geach (ed.) *Logical Investigations*, Blackwell, Oxford.
Gasking, D.: 1996, 'Causation and Recipes', in *Language, Logic and Causation* edited by T. Oakley and L. O'Neill, Melbourne University Press, Melbourne (first published in *Mind*, 1955).
Hacking, I.: 1983, *Representing and Intervening*, Cambridge University Press, Cambridge.
Hellman, G.: 1983, 'Realist Principles', *Philosophy of Science* 50, 227-249.
Howson, C.: 2000, *Hume's Problem: Induction and the Justification of Belief*, Clarendon, Oxford.
Kroon, F. and Nola, R.: 2001, 'Ramsification, Reference Fixing and Incommensurability', in P. Hoyningen-Huene and H. Sankey (eds.) *Incommensurability and Related Matters*, Kluwer Academic Publishers, Dordrecht.
Lewis, P.: 2001, 'Why the Pessimistic Induction is a Fallacy', *Synthese* 129, 371-80.
Musgrave, A.: 1988, 'The Ultimate Argument for Scientific Realism' in R. Nola (ed.) *Relativism and Realism in Science*, Kluwer, Dordrecht, pp. 229-252.
Niiniluoto, I.: 1999, *Critical Scientific Realism*, Clarendon, Oxford.
Norton, J.: 2000, 'How We Know About Electrons', in R. Nola and H. Sankey (eds.) *After Popper, Kuhn and Feyerabend: Recent Issues in Theories of Scientific Method*, Kluwer, Dordrecht.
Peirce, C.: 1901, 'Abduction and Induction', reprinted in J. Buchler (ed.) *Philosophical Writings of Peirce*, Dover, New York, 1955.
Psillos, S.: 1996, 'On van Fraassen's Critique of Adductive Reasoning', *The Philosophical Quarterly* 46, 31-47.
Psillos, S.: 1997, 'How Not to Defend Constructive Empiricism: A Rejoinder', *The Philosophical Quarterly* 47, 369-72.
Putnam, H.: 1978, *Meaning and the Moral Sciences*, Routledge and Kegan Paul, London.
Rescher, N.: 2000, 'Induction as Enthymematic' (unpublished).
Semmelweis, I.: 1983, *The Etiology, Concept and Prophylaxis of Childbed Fever*, translated with introduction by K. Codell Carter, The University of Wisconsin Press, Wisconsin.
Shimony, A.: 1993, 'Scientific Inference' in *Search for a Naturalistic World View, Volume I: Scientific Methodology and Epistemology*, Cambridge University Press, Cambridge.
Tooley, M.: 1987, *Causation: A Realist Approach*, Clarendon, Oxford.
Van Fraassen, B.: 1989, *Laws and Symmetry*, Clarendon, Oxford.
Will, C.: 1995, *Was Einstein Right?*, Oxford University Press, Oxford, second edition.

NOTES

[1]See Ellis (1979) p. 28 footnote 15; Devitt (1997), Chapter 2; Hacking (1983), p. 27 for his distinction between entity and theory realism (this latter being defined in terms of truth); Cartwright (1983), p. 7 for the role of causal reasoning in establishing beliefs about the existence of theoretical entities (but not beliefs about the truth of most theoretical claims such as laws); Hellman (1983), Section 2, for his 'purely ontological formulation' of realism.
[2]For a good introduction to Semmelweis' work, and his subsequent book, see both Carter's introduction to, and his edition of, Semmelweis' one and only book, Semmelweis (1983).

[3]See Kroon and Nola (2001) for an account of reference fixing that can take place even in the presence of false beliefs, and allows for much change in belief while there is continuity of reference, especially in the case of theory change. See also Norton (2000) for an account of continuity in reference to electrons through various changing phases of Bohr's theory of electrons.

[4]The pessimistic meta-induction is an unsound and fallacious statistical inference. What needs to be carried out for an even satisfactory inference is an examination of randomly chosen pairs of theories (separated by some time interval) in some domain to examine the extent to which their central theoretical terms have changed their reference. On even a cursory examination the continuities of reference far outweigh the discontinuities. That it is unsound and fallacious is also argued on different grounds in Lewis (2001).

[5]On this point and others critical of Gasking's analysis see Tooley (1987) Chapter 7.6.

[6]For examples of such experiments, consider, first, the apparatus originally built to detect gravity waves. Initially it was thought that some had been detected. But then it was argued that the detectors originally build by Weber, whatever they responded to, simply failed to detect any gravity waves; his detector failed to reach the required degree of sensitivity. For a readable account of this failure see Will (1995) pp. 253-7. Again the experiments set up in mines thousands of feet below the Earth's surface to detect solar neutrinos emitted from the centre of the Sun failed to detect them in the right amounts. Either the detector was missing many neutrinos, or the theory of the Sun was defective in some respects and the Sun theory predicted that too many neutrinos would be emitted from its centre, or the theory of neutrinos needed to be modified. For an excellent account of this see Franklin (2001), especially Chapter 8 on 'The Missing Solar Neutrinos'.

[7]See Cartwright (1983) pages 6, 85, 92 and 94, and the discussion on p. 92, which is about the spraying of electrons and positrons as described above. A similar position distinguishing IPC from IBE is argued in Clarke (2001).

[8]It is assumed throughout this paper that there is some principled way of drawing a distinction between what might be called 'O-terms', that is observational terms or old or ordinary terms that have their meaning fixed in ways that bring them close to what we can observe, and 'T-terms' which are about unobservable or theoretical items. Though the distinction has come under heavy criticism, some such distinction is needed in order to draw a distinction between theories which are about unobservables and their 'positivistic' versions which are not.

[9]Though it is clear that many non-realists are not positivists, here the distinction does not matter and sometimes the term 'positivist' will also do duty for 'non-realist'.

[10]For van Fraassen's incoherence objections, and his construction of a 'new' epistemology that eschews anything like the traditional IBE, see his (1989) Chapter 7, especially Section 7.4. For other criticisms and defences see, for example, Day and Kincaid (1994), Ben-Menahem (1990) and Psillos (1996) and his reply to his critics in Psillos (1997).

[11]This result appears in an unpublished paper, dated May 18, 2000, by Nicolas Rescher entitled 'Induction as Enthymematic'.

[12]There are many forms of IBE to consider not all of which are seriously defective; for example see the deductivist version of IBE in Musgrave (1988); or those which infer not to the truth of a theory but its verisimilitude (see Niiniluoto 1999, Chapter 6); or Norton's 'demonstrative induction' (Norton 2000); or the defences mounted in Ben Menahem (1990) and Day and Kincaid (1994).

STEVE CLARKE

CHALMERS' DEFENCE OF SCIENTIFIC REALISM AGAINST CARTWRIGHT

1. INTRODUCTION

At least since the publication of her *How the Laws of Physics Lie* (1983), Nancy Cartwright has had a reputation as one of the most significant opponents of scientific realism. A testament to the significance of her work, for scientific realists, is the energy that one particular scientific realist, Alan Chalmers, has devoted to attempting to deflect the threat posed by Cartwright, having criticised her in no less than six articles, published over a thirteen year period (Chalmers 1987; 1988; 1993; 1996; 1999a and 1999b).[1] In this paper I examine Chalmers' efforts to defend scientific realism from the challenges that Cartwright poses. I discern two forms of argument that he employs. These are (1) attempting to argue for the compatibility of Cartwright's observations about scientific explanatory practice with scientific realism, and (2) directly attacking the coherence of Cartwright's alternative to scientific realism. I show that the former is insufficient to deflect the threat that Cartwright presents to scientific realism and the latter is unsuccessful.

The fact that one realist has failed to develop a satisfactory response to Cartwright is not, in itself, overly significant. However, it becomes more significant when we consider that Chalmers has been a leading figure in the philosophy of science for three decades. Furthermore, it can be reasonably assumed, given the determination that Chalmers has displayed in his opposition to Cartwright, that he has apprised himself of the various arguments that have been mounted against her by others. If he believed that any of these were efficacious he would presumably have used these himself, or at least mentioned them, in his many writings on the subject. The fact that a leading scientific realist philosopher has failed to provide a satisfactory response to Cartwright, despite many years of trying, is *prima facie* evidence that there is not a good response to her, that is currently

S. Clarke and T.D. Lyons (eds.), Recent Themes in the Philosophy of Science, 25–38.

available to the scientific realist, and that is a significant conclusion to draw.[2]

Debates between Cartwright and her scientific realist opponents are somewhat complicated by the way in which Cartwright's position has been further articulated after the publication of *How the Laws of Physics Lie* (1983), the book that motivated Chalmers and other realists to set themselves against her. I begin by outlining Cartwright's 1983 position, which inspired Chalmers' opposition. I then identify two important ways in which it has been subsequently articulated by her, that are relevant to discussion of Chalmers' criticisms.

2. CARTWRIGHT'S DEVELOPING POSITION AND ITS CHALLENGE TO SCIENTIFIC REALISM

Despite its name, Cartwright's *How the Laws of Physics Lie* (1983) is not an argument for one simple conclusion. It is a work that brings together arguments for three interrelated conclusions. These are:

> (1) The manifest explanatory power of fundamental laws does not argue for their truth.
>
> (2) In fact the way they are used in explanation argues for their falsehood. We explain by *ceteris paribus* laws, by composition of causes, and by approximations that improve on what the fundamental laws dictate. In all of these cases the fundamental laws patently do not get the facts right.
>
> (3) The appearance of truth comes from a bad model of explanation, a model that ties laws directly to reality. (Cartwright 1983, pp. 3-4)

These striking claims are backed up by a detailed examination of the ways in which fundamental laws are used in actual scientific explanations. In contrast to the low-level phenomenological laws of science that are intended to accurately describe the empirical realm, and about which Cartwright is a realist, fundamental laws are understood by her to be generalisations that are true of idealized models, but are not even approximately true of the world itself. We cannot infer that laws, which are true of the idealized models of science, are even approximately true of the world because of the unsystematic nature of the relations between the idealized models of science and actual reality. Papineau brings out the importance of this point in a summary of Cartwright's case against realism about fundamental laws:

> Cartwright argues that the standard 'derivations' of (physical) effects from fundamental theory are mediated by *ad hoc* assumptions, mathematical short cuts, and fudge factors. Since these devices are in general unmotivated by the fundamental theory, the derivation fails to provide any inductive support for the fundamental theory. The basic theory effectively does no work in the derivations, by comparison with the simplifications, argues Cartwright, and so deserves no credit. (Papineau 1996, p.19)

Scientific Realists believe in the truth, or approximate truth, of the theories that scientists tell them are the best explanations of natural phenomena. Scientific realists argue that, if an explanation is the best explanation by the lights of science, then we are warranted in believing that it is true or approximately true. Furthermore, scientific realists advance the 'success of science argument', which has it that realism is the best explanation for the empirical success of science. If the success of science argument is not made then realism cannot receive the credit for any successes which we might be warranted in attributing to science.[3] So, scientific realists rely crucially on the viability of inference to the best explanation in at least two ways to make their case.

The viability of inference to the best explanation is at the heart of debates between scientific realists and contemporary empiricists. Notoriously, van Fraassen (1983) disputes that being the best explanation need have anything much to do with being true. Scientific theories are selected by scientists as 'the best', in part, because of their non-empirical virtues, such as elegance and simplicity.[4] These are considerations of their utility to us, not considerations that are obviously relevant to an assessment of their closeness to truth.

The challenge of Cartwright's *How the Laws of Physics Lie* (1983), for scientific realists, is twofold. First, it advances a way of thinking about scientific explanation that is very different from the ways that realists favour, and which is amenable to antirealist conclusions. This is the 'simulacrum view' in which scientific explanations are understood as explanations of idealized simulacra that are unsystematically related to reality. The scientific realist needs either to show why this approach to actual scientific explanation is mistaken, or show how, despite appearances, the evidence that Cartwright marshalls in favour of the simulacrum view of explanation can be accommodated by scientific realists. As we will see, Chalmers attempts to take the latter course of action.

Second, the alternative way of accounting for the explanatory role of fundamental laws and scientific theories, described in *How the Laws of Physics Lie* (1983), deepens the threat to the viability of both the inference that our best scientific theories are true, or approximately true, and the inference that realism is the best explanation of science's success, due to van Fraassen. Now scientific realists are faced with a fleshed out alternative way of thinking about what follows from being the best explanation. Scientific realists claim that our best fundamental scientific explanations describe truths about the world, at least approximately. Cartwright tells us that the fundamental explanations that appear to us to be the best explanations of natural phenomena are actually explanations of behaviour that would take place in idealized simulacrums of the world that are easier to explain then messy reality. On her view our best explanations are explanations that misrepresent the actual world. The scientific realist needs to show us why the former account of the explanatory success of the theories that have the fundamental laws of physics at their heart is more plausible than the latter.

Cartwright's ideas have continued to evolve since the publication of *How the Laws of Physics Lie* (1983). Before going on to consider Chalmers' responses to Cartwright we need to take note of two particular developments. In *Nature's Capacities and Their Measurement* (1989) Cartwright advocates realism about causal capacities. She had already advocated realism about causes, as well as about phenomenological laws, alongside theoretical antirealism (Cartwright 1983), so this was not, on the face of it, a major change of position. However, it needs mentioning, because it is the basis of Chalmers' (1993) challenge to the coherence of her position.

In her subsequent writings, Cartwright has become more explicit about the metaphysical consequences of her position (Cartwright 1994; 1999). She now describes herself as an advocate of metaphysical nomological pluralism, the view that 'nature is governed in different domains by different systems of laws not necessarily related to each other in any systematic or uniform way; by a patchwork of laws' (1999, p. 31). This view of the underlying disunity of nature is compatible with an apparent unity of fundamental explanations in science, because, as we have seen, for Cartwright, fundamental scientific explanations are not direct representations of reality. The advocacy of metaphysical nomological pluralism makes Cartwright's opposition to the scientific realist's success of science argument all the clearer. If reality has an underlying disunity, not reflected in scientific explanation, then scientific explanation does not succeed by representing reality as it is.

3. CHALMERS' EVOLVING RESPONSE TO CARTWRIGHT

Phase One: 1987-8, Accommodation

Alan Chalmers' first published discussions of Cartwright's ideas occur in two papers that appeared in the late 1980s (Chalmers 1987; 1988). In both of these papers Chalmers takes note of Cartwright's discussion of laws of nature and argues that the points she makes about the roles that the laws of physics play in actual scientific explanation can be accommodated by the metaphysically rich 'transcendental realism' developed by Roy Bhaskar. He informs us that 'Cartwright's examples describe situations that are entirely to be expected if Bhaskar's realism is correct and so do not constitute a case against it' (1987, p. 87). Transcendental realists advocate realism about causal powers (as well as tendencies and 'generative mechanisms') and understand laws of nature to be generalisations about the invariant behaviour of causal powers in the absence of interference. This is a different view of the ontological status of laws of nature from the anti-metaphysical Humean views that Cartwright had been concerned to attack in *How the Laws of Physics Lie* (1983). Whereas Humeans have extreme difficulty accommodating differences between generalizations about raw empirical data and the generalisations that scientists make when they identify laws of nature, transcendental realists have no such difficulty. Because of its metaphysical richness, it is plausible to hold that transcendental realism can accommodate Cartwright's insights about scientific explanatory practice.

Unlike scientific realists who appeal to a success of science argument to justify realism about explanatorily successful scientific laws, Bhaskar mounts a transcendental argument for realism, which is why he is a self-described 'transcendental realist'. We are warranted in believing that the fundamental laws of science are true, according to Bhaskar, if and only if we make a transcendental presupposition that the causal powers that scientists identify in laboratory experiments — in 'closed systems' in his terminology — are present in the broader world — present in 'open systems', in his terminology — even if their presence is not apparent in the complexity and confusion of open systems. Making this transcendental presupposition is the *only* way to render scientific practice intelligible, according to Bhaskar (1975), and as scientific practice is intelligible to us we have no choice but to make it, or so Bhaskar assures us. It is difficult to overstress the importance of this argument for Bhaskar. As one sympathetic commentator puts it: 'Most of the leading ideas of transcendental realism are rooted in a single transcendental argument which answers the question "how are experiments possible?"' (Collier 1994, p. 31).

Bhaskar's transcendental realism has the potential to provide responses to both of the key problems that *How the Laws of Physics Lie* (1983) raises for scientific realism. It appears that adopting transcendental realism enables us to accommodate Cartwright's observations about the explanatory roles that the fundamental laws of physics play and enables us to avoid relying on a success of science argument. However, very few philosophers have been willing to embrace transcendental realism. Chalmers is no exception. Despite evincing considerable enthusiasm for Bhaskar's ontology, he makes it clear that, in the end, he is not willing to accept Bhaskar's key transcendental argument. Chalmers believes, contra Bhaskar, that 'the possibility of rival accounts of science equally able to render experimental activity etc. intelligible cannot be ruled out' (1988, p. 19). In the place of a transcendental argument, Chalmers urges that we appeal to the 'highly successful' nature of science to motivate realism about the fundamental laws of nature (1988, p. 19).

Chalmers commitment is to a metaphysically rich scientific realism, and not to transcendental realism.[5] A metaphysically rich, but non-transcendental realism does not evade reliance on the success of science argument. Chalmers may be able to accommodate Cartwright's observations about the explanatory roles of the fundamental laws of physics, but showing that both realists and theoretical antirealists are able to explain the same data is not yet to provide a reason to prefer realism over theoretical antirealism. Because scientific realists claim that realism is the best explanation of the success of science, they need to show why realism is the only explanation of the success of science, or alternatively, show that realism provides an explanation of the success of science, which is more likely to be true than the alternatives that theoretical antirealists, such as Cartwright, provide. But Chalmers has done neither of these things.

Phase Two: 1993-6, Attack

Chalmers aimed to push Cartwright 'in the direction of Bhaskar's version of realism' (1987, p. 82). With the publication of *Nature's Capacities and Their Measurement* (Cartwright 1989), it may have appeared to him that he had succeeded. In that book Cartwright embraced realism about causal capacities, in addition to the realism about singular causes and phenomenological laws that had characterised *How the Laws of Physics Lie* (1983). Like Bhaskar, Cartwright agreed that explanatory exportation from experimental situations to the 'open systems' of uncontrolled nature takes place in science. Unlike Bhaskar, she did not see the need for realism about causal capacities outside of those closed systems, as a requirement for us to make sense of this practice. Cartwright did not retract her earlier claim that the fundamental laws of physics were best understood as being false; in fact she reaffirmed this position.

Chalmers' response to the development of Cartwright's position was to accuse her of holding an apparently contradictory position (Chalmers 1993). If capacities are what make laws true then Cartwright, having accepted causal capacities, should be forced, by considerations of consistency, to allow that laws about capacities are true, or so Chalmers asserted. If Chalmers' charge that Cartwright's position was contradictory was upheld, then it was surely not a genuine alternative to scientific realism. And if it could be shown that there is no genuine alternative to scientific realism, then the success of science argument for scientific realism might yet be sustained.

In a response to Chalmers, I pointed out that there was a natural way to read Cartwright, which removed the appearance of contradiction from her work (Clarke 1995). Chalmers interpreted Cartwright as being committed to the view that fundamental laws are generalisations that we accept because we accept that causal capacities are generally present in the magnitudes which we apparently attribute them to the world when we accept laws. However, we can interpret Cartwright as holding that fundamental laws are 'licenses to export' explanations from situations where we understand how capacities behave, to other situations where we do not. The situations where we understand how causal capacities behave are typically simple situations, which we construct in laboratories. The situations where we were able to misrepresent the generality of capacities are typically complex situations where multiplicities of capacities are present. There is no good reason to suppose that we only license ourselves to export claims that are true of the world in general, or so I argued.

Chalmers responded to my defence of Cartwright, conceding that the interpretation of Cartwright that I offered was more plausible than the interpretation he offered, because it had the virtue of removing the appearance of contradiction and because it cohered with Cartwright's other published work (Chalmers 1996). He then retreated to a new version of the accommodation argument that he had developed earlier, and which we have already considered, urging that the fundamental laws of physics are 'strong candidates for the truth' (1996, p. 152). He also argued for the 'more restricted claim' that 'an interpretation of fundamental laws as candidates for true descriptions of the world is implicit in the practice of physics' (1996, p. 152).

Chalmers is presumably right that physicists proceed as if the laws they identify in controlled laboratory conditions are laws that govern the behaviour of the world generally. However, this is not in itself an adequate reason to believe that such laws really do govern the behaviour of the world generally. Cartwright's articulation of metaphysical nomological pluralism is useful in helping to make this point clear. The simple world portrayed by adopting realism about the fundamental explanatory laws of physics may be regarded as an explanatorily useful fiction that is only true of some aspects of our complex, messy reality. Once again the stumbling block for scientific realists is an inability to make the success of science argument go through. Chalmers wishes to infer that because the explanatory success of fundamental laws is accounted for by adopting realism about such laws, realism is likely to be true. Cartwright provides us with an alternative framework for understanding the explanatory success of the fundamental laws of physics, in which they need not be true. Chalmers wishes to sustain the claim that the former explanation of science's explanatory success is somehow better than the latter, but fails to demonstrate that this is so.[6]

Phase Three: 1999, Recent Developments

In a broad-ranging recent discussion of the place of laws of physics, Chalmers points out how dissatisfying Cartwright's philosophy is to those who are possessed of the desire to explain the behaviour of nature (Chalmers 1999a). In his words:

> Once we follow Cartwright and deny a descriptive role to fundamental laws, then we run into the same problems as beset the regularity view that restricts the applicability of laws to those experimental situations where the appropriate regularities obtain. We are at a loss to say what governs the world outside of experimental situations. (Chalmers 1999a, p. 9)

Chalmers is right that this situation feels dissatisfying. But how are we to respond to this dissatisfaction? The empiricist outlook that Cartwright shares with van Fraassen is one in which we are continually urged to relate our beliefs to evidence and to be suspicious of the motives that drive us to prefer particular explanations over others. Empiricists distrust inferences to the best explanation because they suspect that some explanations end up being judged as being better than others in virtue of being more satisfying to us, rather than being more plausibly true. Chalmers urges us to respond to the dissatisfying endpoint that Cartwright arrives at by rejecting Cartwright's philosophy. Cartwright, however, would urge us to come to accept our descriptive limitations and make peace with them.

In a second paper, also published in 1999, Chalmers contemplates, but stops short of endorsing, a transcendental argument for the conclusion that 'laws identified in experimental contexts are presumed to apply outside of those contexts' (Chalmers 1999b, p. 337). He contends that 'This move, that only renders explicit what is implicit in science and its application, neutralises *all* of the arguments that Nancy Cartwright (1983) has put forward as reasons for claiming that the laws of physics lie' (1999b, p. 337). We have come full circle. The argument that Chalmers contemplates is of course Bhaskar's key argument for transcendental realism, which Chalmers stopped short of embracing in his earliest writings on Cartwright and which he still flirts with, but fails to embrace.

4. CHALMERS' REALISM AND SCIENTIFIC REALISM

I have characterised Chalmers as a defender of scientific realism, and one who has been particularly concerned to defend scientific realism from the challenge to it posed by Cartwright. Chalmers makes it clear, whenever his writings broach metaphysical topics, that he wishes to be regarded as a realist. However, he appears not to regard himself as a *scientific* realist (1999c, p. 238). Have I misrepresented Chalmers' opposition to Cartwright by portraying it as a defence of a position that Chalmers does not advocate? In this section of the paper I examine Chalmers' views about realism more closely and show that he is, in fact, a scientific realist, in as much as he is committed to key scientific realist doctrines.

Chalmers' most sustained discussion of realism appears in a new chapter of the recently revised third edition of his *What is this Thing Called Science?* (1999). Here he characterises scientific realism in a way that is broadly in agreement with the characterisation presented in the introduction to this volume. Lyons and Clarke (this volume, p. ix) identify two key commitments made by advocates of scientific realism. A commitment to an accurate representation of the world as the core aim of science, and a commitment to the view that science succeeds in achieving this aim, at least approximately. Lyons and Clarke go on to note the key role of the 'success of science' argument in securing the credit on behalf of realism for science's empirical success (this volume, pp. xi-xiii). The best explanation of the success of science is that science has achieved or approximated its aim, that the world really is as science describes it, or so says the scientific realist.

Chalmers (1999c) appears to be in substantial agreement with Lyons and Clarke's characterisation of the aim of science, according to scientific realism. In his words: 'According to scientific realism, science aims at true statements about what there is in the world and how it behaves, at all levels …' (1999c, p. 238). He also agrees with Lyons and Clarke (this volume, *passim*) about the typical scientific realists' reliance on a notion of approximation to truth in attempting to demonstrate how science can achieve this aim: 'We cannot know that our current theories are true, but they are truer than earlier theories, and will retain at least approximate truth when they are replaced by something more accurate in the near future.' (1999c, p. 238) Like Lyons and Clarke, he also takes note of the importance of the 'success of science' argument in securing, for scientific realism, the credit for science's successes: 'It is claimed that scientific realism is the best explanation of the success of science and can be tested against the history of science and contemporary science in much the same way as scientific theories are tested against the world.' (1999c, p. 238)

Although Chalmers (1999c) provides a quite detailed characterisation of scientific realism, he stops short of endorsing this 'very strong form' of realism (1999c, p. 238), because he worries about the force of some antirealists arguments against it, most notably the argument from the 'dismal induction'. This is an argument from the observation that successive theories, that scientists have judged to be successful in the past, are now believed to be false. Its proponents urge us to accept an inductive argument for the antirealist conclusion that our current scientific theories are probably false too.

A standard scientific realist response to the dismal induction is to argue that the theories that scientists adhered to in the past, and which they now judge to be false, are not out-and-out false. Rather they are approximately true. Unfortunately, scientific realists have found it exceedingly difficult to articulate exactly what they mean when they say that a theory is approximately true.[7] Chalmers, in response to the threat of the dismal induction, also appeals to the notion of approximate truth. However, in the process of doing this he understands himself to be retreating from scientific realism to a weaker realist position. He refers to this position as 'unrepresentative realism', which he describes as being very similar to John Worrall's 'structural realism' (Worrall 1989). Unrepresentative realists claim that science aims at, and succeeds in, describing approximate truths about the world, but only insofar as it produces mathematical models that aim at representing the structure of reality. In Chalmers' words:

> …science is realist to the extent that it attempts to characterise the structure of reality, and has made steady progress insofar as it has succeeded in doing so to an increasingly accurate degree. Past scientific theories were predictively successful to the extent that they did at least approximately capture the structure of reality. (Chalmers 1999c, p. 245)

I do not dispute Chalmers' claim that unrepresentative realism offers a promising response to the dismal induction. However I do dispute Chalmers' characterisation of unrepresentative realism as a distinct form of realism from scientific realism. Unrepresentative realism, as Chalmers describes it, involves acceptance of all of the key points in Chalmers' characterisation of scientific realism. So, by Chalmers' own lights, it should be seen as a form of scientific realism, rather than an alternative. Scientific realists, as Chalmers informs us, hold that science aims at a true description of the world. The unrepresentative realist agrees, and argues that it is when identifying structural features of the world, which can be mathematically modelled, that science is able to achieve this aim. Scientific realists, as Chalmers informs us, typically argue that we need to appeal to a conception of approximation to truth. The unrepresentative realist agrees and argues that it is in the successive refinement of mathematical models that we need to look to find approximation to truth taking place in science. The scientific realist, as Chalmers notes, utilises a 'success of science' argument. The unrepresentative realist also needs to appeal to a success of science argument. Unrepresentative realists explain the success of science in terms of the ability of mathematical models to represent the world accurately and argue that success is best accounted for by realism. So, on all three of Chalmers' discernible criteria for allegiance to scientific realism, the unrepresentative realist is a scientific realist.

I have argued that unrepresentative realists, such as Chalmers, ought to be characterised as a subspecies of scientific realists rather than as a distinct species. But I have not done this because of a strong concern about the taxonomy of realisms. Both unrepresentative realists and 'strong' scientific realists are committed to core doctrines that Cartwright challenges. As this paper has been concerned to evaluate Chalmers' responses to Cartwright's challenges to these core doctrines, it is the commitment of Chalmers to core scientific realist doctrines that it has been important to establish.

5. CONCLUSION

As we have seen Cartwright (1983) challenges scientific realism in two ways. She advances an account of explanation that is amenable to antirealist conclusions. Furthermore, her acount of the explanatory role of the fundamental laws of physics strengthens antirealist arguments against the inference that our best scientific theories are true or approximately true and against the inference that realism is the best explanation of science's success. In the six papers that we have examined, in which Chalmers has responded to Cartwright's challenge, two forms of argument have been discerned. Chalmers' strong form of argument has been to charge Cartwright with contradiction. As we saw this charge was not sustained and has now been withdrawn. Chalmers' weaker form of argument has been to argue for the compatibility of Cartwright's observations about scientific explanatory practice with scientific realism. As I have now shown, this argument is insufficient to do the work that Chalmers requires of it. Because of scientific realism's reliance on the success of science argument, the demonstration of mere compatibility with Cartwright's observations about scientific explanatory practice is an inadequate defence of scientific realism. Scientific realists need to demonstrate not simply that realism is a possible explanation of scientific explanatory practice, but that realism is the best explanation of scientific explanatory practice and the one we should accept as correct.

In three of the six papers examined, Chalmers has carried on a flirtation with transcendental realism, without being willing to commit himself to it. In my view this reluctance to commit has prevented him from taking a clear stance against Cartwright. As a scientific realist, he cannot expect to be able to fully meet the challenge that Cartwright presents without providing a defence of the success of science argument and of inference to the best explanation in general. Transcendental realism provides a very different route to a response to Cartwright. To appeal to a transcendental argument is to make an appeal to a form of argument that is unknown in science and considered to be illegitimate by the majority of contemporary empiricist philosophers. Clearly embracing transcendental realism would do away with a problem that Chalmers has failed to adequately address, but it would involve facing up to a very different set of problems.[8]

Centre for Applied Philosophy and Public Ethics,
Charles Sturt University

REFERENCES

Bhaskar, R.: 1975, *A Realist Theory of Science*, Leeds Books, Leeds.

Cartwright, N.: 1983, *How the Laws of Physics Lie*, Oxford University Press, Oxford.

Cartwright, N.: 1989, *Nature's Capacities and their Measurement*, Oxford University Press, Oxford.

Cartwright, N.: 1994, 'Fundamentalism Vs. The Patchwork Model of Laws', *Proceedings of the Aristotelian Society* 103, 279-292.

Cartwright, N.: 1999, *The Dappled World*, Cambridge University Press, Cambridge.

Chalmers, A.: 1987, 'Bhaskar, Cartwright and Realism in Physics', *Methodology and Science* 20, 77-96.

Chalmers, A.: 1988, 'Is Bhaskar's Realism Realistic?', *Radical Philosophy* 49, 18-23.
Chalmers, A.: 1988b, 'Realism in Physics: A Reply to Wal Suchting', *Methodology and Science* 21, 1988, 296-9.
Chalmers, A.: 1993, 'So the Laws of Physics Needn't Lie', *Australasian Journal of Philosophy* 71, 196-205.
Chalmers, A.: 1996, 'Cartwright on Fundamental Laws: A Response to Clarke', *Australasian Journal of Philosophy* 74, 150-152.
Chalmers, A.: 1999a, 'Making Sense of Laws of Physics', in H. Sankey (ed) *Causation and Laws of Nature*, Kluwer, Dordrecht, pp. 3-16.
Chalmers A.: 1999b, 'Twenty Years On: Adding the Cat's Whiskers', *Science and Education* 8, 327-338.
Chalmers, A.: 1999c, *What is This Thing Called Science?*, Third Edition, University of Queensland Press, St. Lucia.
Clarke, S.: 1995, 'The Lies Remain the Same: A Reply to Chalmers', *Australasian Journal of Philosophy* 73, 152-155.
Clarke, S.: 1998, *Metaphysics and the Disunity of Scientific Knowledge*, Ashgate, Aldershot.
Clarke, S.: 2001, 'Defensible Territory for Entity Realism', *British Journal for the Philosophy of Science* 52, 701-722.
Collier, A.: 1994, *Critical Realism: An Introduction to Roy Bhaskar's Philosophy*, Verso, London.
Ellis, B.: 2001, *Scientific Essentialism*, Cambridge University Press, Cambridge.
Niiniluoto, I.: 1978, 'Truthlikeness: Comments on Recent Discussions', *Synthese* 38, 281-329.
Papineau, D.: 1996, 'Introduction', in D. Papineau (ed), *The Philosophy of Science*, Oxford University Press, Oxford, pp. 1-20.
Psillos, S.: 1999, *Scientific Realism: How Science Tracks Truth*, Routledge, London.
Putnam, H.: 1975, *Mathematics, Matter and Method*, Cambridge University Press, Cambridge.
Van Fraassen, B.C.: 1980, *The Scientific Image*, Oxford University Press, Oxford.
Van Fraassen, B.C.: 1989, *Laws and Symmetry*, Oxford University Press, Oxford.
Worrall, J.: 1989, 'Structural Realism, the Best of Both Worlds?', *Dialectica* 43, 99-124.

NOTES

[1]I will not comment on a seventh article by Chalmers (1988b), which is only tangentially about Cartwright and does not appear to add any weight to his case against her.

[2]Despite this situation, Cartwright's 'entity realist' alternative to scientific realism is widely believed to be unsustainable. I offer a partial defence of entity realism in Clarke (2001).

[3]A strong form of the success of science argument is the 'no miracles argument' which has it that metaphysical realism provides the best explanation of the success of science because it provides the only explanation that does not render this success miraculous (Putnam 1975).

[4]Cartwright would want to credit Duhem rather than van Fraassen for this line of attack on inference to the best explanation. Van Fraassen also mounts a variety of other lines of attack on inference to the best explanation (1980; 1989).

[5]Chalmers has confirmed this to me, in conversation, describing his metaphysical views as having more in common with those of Brian Ellis than Bhaskar. See, Ellis (2001).

[6]Chalmers' 'Response to Clarke' (Chalmers 1996) is also discussed, in a rather different context, in Clarke (1998, pp. 48-9).
[7]Sustained attempts to articulate a notion of approximate truth took place in the 1970s. See Niiniluoto (1978). More recent developments are discussed by Psillos (1999, pp. 261-279).
[8]Thanks to Brian Ellis and Tim Lyons for helpful comments.

HAROLD KINCAID

SCIENTIFIC REALISM AND THE EMPIRICAL NATURE OF METHODOLOGY: BAYESIANS, ERROR STATISTICIANS AND STATISTICAL INFERENCE

1. INTRODUCTION

Current debates over scientific realism have two suspicious traits: they are about entire domains of modern experimental science and they propose to determine the epistemic status of these domains on the basis of philosophical arguments. The goal of this paper is to show that these assumptions are indeed suspicious and in the process to defend an alternative view — call it contextualist realism — that is not committed to either claim.

Contextualist realism holds that there are no *a priori* universal inference rules that suffice by themselves to evaluate the realism question over entire domains of scientific practice. In positive terms this means that realism questions are local and empirical in a way that makes them strongly continuous with the sciences themselves. The 'realism' in 'contextualist realism' comes from the claim that the local evidence sometimes shows that pieces of science are true, approximately true, successfully refer, well confirmed, or whatever is one's preferred realist idiom. [1]

This paper argues for a realism of this sort inductively by examining two such universal inference rules — Bayesian and error-statistical — and arguing that they depend on empirical evidence in ways unrecognized. What results is some evidence for the contextualist position I advocate.

The paper proceeds as follows: Section 1 briefly outlines contextualist realism and then details the suspicious assumptions identified above as they surface in various defences of realism. Section II then outlines the dispute between the Bayesian and error-statistical approaches, which is both a dispute about general rules of inference and about the foundation of statistics. The Bayesian approach is the subject of Section III. I argue there that Bayesian inference requires empirical input in ways not recognized by

S. Clarke and T.D. Lyons (eds.), Recent Themes in the Philosophy of Science, 39–62.

Bayesians and in ways that error-statistical approaches can acknowledge. Section IV makes similar claims about the error-statistical approach. The end result is a partial defence of the local realist approach, which reasons to dissolve a long standing controversy over statistical inference.

2. CONTEXTUALIST REALISM AND THE LOGIC OF INFERENCE

In this section I outline the contextualist realism I favor and explain how difficulties I identify for Bayesian and error-statistical approaches constitute evidence on its behalf. I begin with a general statement of contextualism about epistemology and follow with what it says about realism and scientific inference.

'Contextualism' as I use the term is a variant of naturalized epistemology. It shares the naturalist claim that epistemological theories cannot be developed without knowing something about our place in the universe and thus without investigating our cognitive abilities with the best empirical tools we have. Put more strongly, empirical facts are not only relevant to epistemology but are all there is — a priori standards of foundationalism have no place. Contextualism adds to this rejection of foundationalism the following claims about how our justificatory practices actually work:

(1) We are never in the situation of evaluating all of our knowledge at once.

(2) Our 'knowledge of the world' is not a coherent kind that is susceptible to uniform theoretical analysis.

(3) There are no global criteria for deciding which beliefs or principles of inference have epistemic priority.

(4) Justification is always relative to a specific context, which is specified by the questions to be answered, the relevant error possibilities to be avoided, the background knowledge that is taken as given, etc.

Applied to the issue of scientific realism, contextualism has negative and positive morals. The negative moral is that arguments for and against scientific realism cannot proceed by evaluating all of science at once, by appealing purely to formal or methodological grounds, or by failing to invoke substantive empirical background information. So *global* realist and antirealist arguments are equally misguided. Put positively, arguments over realism must proceed by assessing specific theories or fragments thereof, given what else is known in the relevant context. Such arguments are as much scientific as philosophical.

To see the upshot of such contextualism, consider the implications for two standard moves in the realism debate, the pessimistic induction and the argument from underdetermination. The pessimistic induction says that current successful science could be in error because we know past successful science has been. The argument from underdetermination says that once we have all the data, multiple competing theories will be supported by the data. On the contextualist view, these arguments are not so much valid or invalid as misguided. In specific scientific contexts where background knowledge and questions to be answered are set, pessimistic inductions and arguments from underdetermination can be debated and realist or antirealist conclusions reached. But there is no coherent project of taking 'all the data' and asking what theories of the world they 'support' in the abstract, or of doing an induction on 'success' and the total past record of scientific inquiry.

With this background on contextualism, let's look more specifically at the versions of scientific realism I reject and at how arguments over Bayesian and error-statistical approaches might support the contextualist alternative I favor. The standard argument for scientific realism appeals to inference to the best explanation in some form. We can only explain the success of modern science, it is claimed, on the assumption that its methods are reliable indicators of the truth. Because its methods apply equally to observational and theoretical claims, we have reason to be realists about both parts of science.

These arguments presuppose the view of realism I want to reject. They are committed implicitly to a 'logic of science' ideal. Let me first be clear on that ideal. Discussions of scientific inference up to the present have hoped to find a logic of science, taking deductive logic as a paradigm. That means any successful account of scientific method and inference must describe rules that at least are jointly:

Universal: The rules for good inference should hold across domains. They should apply regardless of the specific science in question or the time at which the science is practiced; they should hold for 19th century physics and 20th century biology, however disparate their subject matter and practitioners.

Formal: The rules for good inference should be about form, not content. They should not depend on specific empirical assumptions.

A priori: The rules for good inference are not justified by empirical evidence. Exactly what sort of non-synthetic truths they are varies from program to program, but some sort of conceptual truth is one standard answer.

Sufficient: Rules of good inference should allow us to infer whether evidence *e* confirms hypothesis *h* over *not h* without the use of further information. The evidence and hypotheses can of course be complex, but once all the relevant data and theory are present, inference should rely on the logic of science alone.

These are no doubt stringent requirements, but as an ideal they have had great influence. Carnap, for example, wanted an account of confirmation that exhibited 'a logical relation between two statements', one 'not dependent on any synthetic statements' (1950, p. v). Modern-day advocates are also common. Campbell and Vinci (1983) argue that the confirmation relation between evidence and hypothesis is analogous to 'the entailment relation in deductive logic.' The same view motivates discussions of bootstrap approaches to confirmation and alleged counterexamples. Likewise, the logic of science ideal is prevalent among statisticians as well. To take one typical example, Efron and Tibshirani propose an 'ideal computer-based statistical inference machine … The statistician enters the data, the questions of interest … Without further intervention, the machine answers the questions' (1993, p. 393). This picture of statistical inference is widespread among practitioners in the social, behavioral, and biomedical sciences.

The hope for a logic of science is best thought of as an ideal that can be approached to varying degrees. Each of the requirements for a logic of scientific inference can be given stronger and weaker readings. Demands for universality can be restricted by some prior intuitions about the domain of good science. Requiring a formal logic can be weakened to the demand that no *domain specific* empirical facts be used. Variations are likewise possible on the claim that methodological rules are *a priori*. Empirical facts might be relevant in different ways and to different degrees. If methodological norms are tested against our prior intuitions about good science, then at least the empirical facts about those intuitions are relevant. The demand that a logic be sufficient can vary according to just how broadly the relevant data and hypothesis are described. The less background information that is folded into *h* and *e*, the stronger the claim. Similarly, we might distinguish rules that are sufficient in the sense of presupposing no other methodological or normative information as opposed to those that require no other information *simpliciter*. Crudely put, the key issue in all these cases concerns how much work one's rules for good inference do.[2]

The first explicit defence of scientific realism as an inference to the best explanation that I know of is Maxwell's: 'as our theoretical knowledge increases in scope and power, the competitors of realism become more and more convoluted and ad hoc … they do not explain why the theories … can make such powerful, successful predictions. Realism explains this very simply . . .' (1970, p. 12). Maxwell cashes this out in explicitly Bayesian terms. Since both realism and antirealism can tell stories about why science is successful, then the ratio of the relevant likelihoods is unity. Thus what distinguishes the views are the relevant priors. Realist explanations are simpler, more comprehensive, and less *ad hoc*. Therefore their prior probabilities are higher and they are thus better explanations.

The logic of science ideal enters twice in this argument: first in appealing to Bayes' theorem as the final arbiter, and second, in invoking simplicity, comprehensiveness, and *ad hoc*ness. Both are treated as universal criteria of theory choice — as standards that have an interpretation and justification independent of any particular scientific domain and results.

More recent scientific realists such as Boyd (1990) explicitly eschew the logic of science ideal. Yet arguably their defences of realism *presuppose* that ideal nonetheless. Boyd directly rejects the local piecemeal approach to realism advocated here. He does so because 'the attraction of scientific realism is that it appears to offer a distinctive and coherent conception of scientific knowledge — one which, for example, preserves a certain common sense ... conception of the way in which scientists exploit causal interactions with natural phenomena to obtain new knowledge' (p. 175). That distinctive approach is based on the claim that a 'realist understanding of scientific theories [is] part of the best naturalistic explanation for the success of various features of scientific method' (p. 180). But no such explanation is forthcoming unless there are universal standards of explanation that suffice to tell us which explanation of science is the best. We are being given a philosophical argument appealing to universal standards to evaluate entire disciplines. The logic of science ideal is not far away.

A particularly vivid illustration of the logic of science ideal is found in Trout's (1998) defence of realism about the social and behavioral sciences. The argument is again based on the success afforded by methods and is again a global defence with similar equivocations. The method Trout cites are the standard statistical practices of the social and behavioral sciences. Those practices are a form of measurement, he argues, and their success can only be explained on the assumption that their theoretical terms measure real, independently existing causal processes. Trout's defence is of social science across the board. Yet he too has to deal with the piecemeal skeptic. He does so by defending moderate realism, which is apparently the view that all and only those parts of social science that result from statistical testing are to be taken realistically.

The presupposition of universal inference rules that suffice to evaluate entire domains is obvious in Trout — it is in the appeal to the rules of good statistical practice. Our reliance on those rules is justified by a philosophical argument, namely, that the best explanation for their success is their reliability in producing true theories. The discussion that follows of Bayesian and error-statistical approaches directly challenges those assumptions.

3. THE DISPUTE BETWEEN BAYESIAN AND ERROR-STATISTICAL APPROACHES

This section outlines the issues in the dispute between Bayesian and error-statistical approaches.[3] They differ both on fundamentals and on the specific inferences they find justified. At the level of fundamentals the differences concern the nature of probability, what probabilities attach to, the role of background beliefs, and more. For Bayesians, all inferences from data to hypotheses, including statistical hypotheses, must involve updating probabilities in accordance with Bayes' theorem. The theorem says that we should evaluate the probability of a hypothesis *h* given evidence *e* by asking whether $p(h/e) > p(h)$. This is done by factoring in the prior probability of *h*, the probability of the evidence if *h* were the case, the probability of the competing hypothesis, *not h*, and the probability of the evidence *e*, if *not h* were true. These parameters are then combined to determine the probability of *h* in light of *e* by Bayes' theorem, one form of which is:

$$p(h/e) = p(h) \times p(e/h) \div p(h) \times p(e/h) + p(not\ h) \times p(e/not\ h)$$

For most Bayesians, the probabilities involved are individual levels of confidence or degree of belief, and they attach to specific statements or hypotheses. Evidence is taken to support a hypothesis when the probability of, or confidence in, H given observed evidence E is greater than the confidence in H without E.

For the error-statistical approach, probability is long run frequency, and thus probabilities do not attach to specific propositions. They can, however, consistently be applied to testing procedures. The error rate of a given test procedure is an objective fact that requires no determination of the experimenter's subjective degree of belief. Confirming a hypothesis means ruling out different kinds of possible error. That is done by subjecting the hypothesis at issue to severe tests. A severe test, on Mayo's construal, is one in which the probability is low that the test will pass *h* — show no error present — when *h* is in fact false. Interpreted correctly, this is supposedly just what classical statistical methods give us. When a hypothesis has been severely tested, then we have good reason to think it reliable or true.

The (alleged) differences between the two approaches become more visible when we look at specific practices, particularly statistical practices. I will mention three here: using known evidence to construct hypotheses compared to novel predictions, stopping rules, and significance testing.

Bayesians typically deny that it matters whether evidence is old or new, used in constructing the hypothesis or not. Their reasoning is that all that matters is the *p(h/e)* and that how and when the evidence was known plays no role in determining the priors and likelihoods needed to evaluate that quantity. Making that information count would require making the precise mental state of an investigator crucial to determining whether the evidence supports an hypothesis, and that seems absurd. On the error-statistical view, evidence used in constructing a hypothesis can count for less if doing so results in a testing procedure that is not severe. It is also easy to construct cases that support this claim — instances where an hypothesis seems to fit the evidence only because it was designed to do so, not because it is likely to be true. This time the error-statistical view looks convincing: this method of testing — confronting *post hoc* hypotheses with the data used to construct them — frequently passes false hypotheses.

A second sharp disagreement between the Bayesian and error-statistical approach concerns 'stopping rules' in statistical testing. A stopping rule specifies the possible set of outcomes in a trial. A significance test uses that set of possible outcomes to determine the probability that the observed outcome of a trial would occur even if the null hypothesis were true. Error statisticians think stopping rules are essential, because they are relevant to determining the error probabilities of a test. Bayesians, however, think stopping rules are irrelevant. Stopping rules seem to imply that whether a particular outcome supports an hypothesis depends on the size — say the number of flips of a coin — of the trial the experimenter intended when she started. It is easy to construct cases where the same actual results in a trial go from significant to insignificant depending on the stopping rule used.[4]

Not surprisingly, significance testing itself is another source of controversy. Aside from its reliance on stopping rules, Bayesians object to significance testing for other reasons. They argue that it is not clear what a decision to 'reject' the null hypothesis comes to. Does it mean the null is probably false? If so, the error-statistical approach is bound to err for two reasons: it draws conclusions about single instances from long run frequencies and does so without factoring in prior probabilities. Long run frequencies are compatible with indefinitely many short-run sequences. So from the fact that a test is reliable in the long run we cannot conclude much about its reliability in this case. Ignoring prior probabilities is likewise troublesome, because it will lead to probabilistic incoherence.

These are, in brief, the issues separating the Bayesians and error statisticians. What they have in common is the logic of science ideal. It is jettisoning that ideal that it is crucial to resolving the dispute between them, or so I argue in the next two sections.

4. THE BAYESIANS

In this section I argue that Bayesian approaches are empirical in ways frequently not recognized and that once we see those empirical components, important parts of the Bayesian/error-statistical dispute in statistics can be resolved. I begin with abstract considerations and then move to concrete examples.

In one sense it is obvious that any Bayesian account makes methodology empirical. After all, part and parcel of the approach is the dependence on priors, and of course those rest on empirical considerations. Bayesians will not deny this in principle. Yet in practice they typically defend the most *a priori* version of their doctrine.

They do so in several ways. First, Bayesians — in response to worries about the subjectivity of their approach — minimize the importance of prior information by arguing that the priors washout: individuals with different priors confronting the same repeated data will eventually come to assign a hypothesis the same posterior probability. There are serious doubts about whether the priors actually do lose their influence (see Earman 1992). So focusing on the situation where the priors do washout, Bayesians are looking for the situation where their doctrine reduces scientific inference to a purely logical matter. More importantly, Bayesians miss and/or deny various constraints on scientific inference that are empirical in nature. As I shall argue next, there are numerous methodological norms that can be reconciled with Bayesianism if those norms are construed as empirical claims that form part of the relevant background knowledge used in applying Bayes' theorem.

Let me first illustrate my claim that Bayesian approaches wrongly leave out important empirical knowledge about methodological norms by discussing a case in point, namely, the place of inference to the best explanation. There is a long tradition of treating inference to the best explanation as a foundational inference rule. Harman (1965), for example, takes it to ground induction. Scientific realists often take it to be the basis for inferring that current science must be approximately true. Yet others like van Fraassen (1980) take it to be incoherent. He envisions calculating *p(h/e)* via Bayes' theorem from priors, and then giving one hypothesis extra credit on the grounds of its explanatory virtues, thus leading to incoherence. Neither view is right.

Taking Harman's claims first, note that our constraints on good explanations will depend on substantive background information that is empirical and defeasible. Moreover, that information will be just one factor among many that go into deciding when a hypothesis is confirmed. So Newton's laws of motion were inferior on explanatory grounds when first proposed in that they violated a fundamental explanatory stricture of the mechanical philosophy, namely, no explanation by occult forces and no action at a distance. But gravity was seen as precisely that. So the explanatory virtues were evaluated via substantive background beliefs. Moreover, those explanatory constraints were eventually overridden by the impressive predictive power of Newtonian mechanics.

Van Fraassen's criticisms turn on treating explanatory virtues as somehow over and above the considerations that go into calculating Bayes' theorem. Yet they need not and should not be so taken. Explanatory constraints are part of determining the priors. For the adherent of the mechanical philosophy, Newton's laws had diminished prior probability to the extent that they apparently invoked a nonmechanical force. So there is no need to add or subtract credit after applying Bayes' theorem. These considerations are involved in applying the theorem itself.

For our purposes, the important point is that the Bayesian approach does not tell us how or what explanatory constraints factor in confirmation. We need to know information about important issues in confirmation theory before it will provide us with answers; some of that information will be in the form of substantive empirical facts, ones that could be domain specific.

So the moral is this: Bayesian accounts of inference are silent on some important methodological controversies. These issues are empirical and ones that must be decided in order to apply Bayes' theorem. In this sense Bayesian inference is empirical in ways not generally acknowledged.[5] The upshot is that Bayesian approaches do not obtain the ideal of a universal, *a priori*, and sufficient inference rule. Of course Bayes' theorem is *a priori* (if anything is), and it may be universal in that all scientific inference must accord with it. Indeed, my arguments have indirectly supported universality by showing that alleged counter instances (e.g. IBE) are no such thing. But this universality comes at the price of content. Bayes' theorem is silent on important methodological issues and thus very far from being sufficient as the logic of science ideal demands.

I want to next generalize these points to the other issues separating the Bayesian and error-statistical approaches. Take first the issue of novelty. Error statisticians believe that the way in which a hypotheses was generated influences its degree of confirmation; the Bayesian view is supposed to be that such information is irrelevant (cf. Mayo, Chapter 9). However, as some Bayesians (Howson and Urbach 1993; Howson and Franklin 1991) and some Bayesian friendly writers have pointed out (Maher 1988), how the data was generated can be evidence that the data was produced by a reliable process and thus can potentially factor in a Bayesian analysis. Maher tries to prove more or less *a priori* (by appeal to very general probability assumptions) that whenever a hypothesis predicts rather than accommodates data, it must be the case that we have evidence that a reliable process is involved. Howson and Franklin grant that there can be such cases, but claim they are rare in science. I want to look at these arguments in a little more detail and argue that (1) error statisticians are clearly wrong that Bayesian accounts have no place for how the data were generated, (2) the import of old versus new evidence is an empirical one that can vary across domains (contra Maher), (3) Howson and Franklin do not see the variety of ways such information can be relevant and are wrong in claiming that novelty has little general relevance in science.

Maher argues that there are good Bayesian reasons to think that novelty of the evidence matters to confirmation. Compare two individuals who predict the outcome of the hundredth toss of a coin, one who has successfully predicted the first 99 and another who has learned the first 99 outcomes. In the first case how the data are generated give us good reason to think the predictor is reliable; in the second case we have no such confidence. Moreover, this situation is like many we find in science. Mendeleev, for example, predicted undiscovered elements, suggesting that he had knowledge of the actual mechanisms at work just as did the reliable coin predictor.

Howson and Franklin grant that predicting novel evidence may matter if it suggests that someone has advanced knowledge about outcomes. If e is the prediction on the first 99 tosses, h is the prediction of the 100th toss plus e, and m is the hypothesis that the subject has advanced knowledge, then

$$p(h/e) = p(h/e\&m)\, p(m/e) + p(h/e\&\text{-}m)\, p(\text{-}m/e)$$

However, $p(m/e)$ will be large by Bayes' theorem, since $p(e/\text{ not } m)$ — that the reliable predictor is simply right by chance — is low and $p(e/m)$ is high. So we have good reason to assign high confidence to $p(h/e)$. In the case of accommodation, $p(e/m)$ is high and $p(e/m)$ is low. So there is a difference between prediction and accommodation in this case. But, Howson and Franklin argue, this will not hold in all cases of novel prediction. In the Mendeleev case, m is really just the claim that the periodic table is true. But m entails e regardless of how Mendeleev came to his predictions. If predicting the new elements had any force, it was only because no other competing theory predicted the data and thus *p(e/not m)* was low. So novelty is not the issue. Actually, the Mendeleev case is quite typical in science, the mysterious coin predictor rare.

I agree with Howson and Franklin that Maher's alleged proof that prediction counts more than accommodation is unconvincing — Maher's project is part of the *a priori* approach that I am rejecting. However, Howson and Franklin's diagnosis seems to me confused. They take *m* in the Mendeleev case to say that *h* is true. But we can equally take it to say the same thing in the case of the coin predictor — to say that the coin predictor's theory of the coin bias is true. Then they should conclude that the coin predictor is likewise no different than the accommodator. Alternatively, we can take *m* in the Mendeleev case to be 'Mendeleev generates predictions by a reliable process.' Then *p(e/m)* is not unity.

The real source of the difference — to the extent that there is one — between the coin predictor and the Mendeleev case is that in the former we have a much better grip on the probability that he got the predictions right by chance. That seems highly unlikely in the coin case. In the Mendeleev case it is less obvious how to assess that possibility. To see this more clearly, redescribe the two cases in a way that depicts the evidence as producing such and such a result in a given kind of test. So *e* now equals predicting the 100 coin tosses correctly by accommodation. In the Mendeleev example, *e* would be predicting gallium after its discovery. Then *p(e/not h)* in both cases is arguably high, though as I said it is more clearly so in the coin example. The chances that one can produce a correct prediction if one accommodates with a false theory are extremely good in the coin case, perhaps greater than .5 in the Mendeleev case. That means of course that *p(h/e)* is going to be correspondingly diminished.

So we are back to the broad theme of the paper. There is no conceptual answer to the question of whether novel data confirm more than accommodated data. Accommodating can be cause for skepticism, particularly when it is easy (as in the coin case). Predicting novel facts need not be a sign of reliability — for example, when the novel fact is predicted after many guesses. There is no purely logical or conceptual way to decide their importance; defences or criticisms of scientific realism that depend on assuming just the opposite are thus misguided.

Thus Mayo is surely wrong to claim that Bayesians can allow no place for novelty. Bayes' theorem will not decide the issue without substantive empirical input, but that does not show that novelty and Baysianism are incompatible. The Bayesian victory here, however, is a bit pyrrhic. For the motivation for incorporating novelty rests on the kind of arguments error statisticians like Mayo think are central to science — those looking at the reliability of various testing procedures. However, that does not mean that

Bayesians have to accept the way error statisticians interpret and use those ideas, and so differences surely remain.

We can now draw similar conclusions about the dispute over stopping rules. The dispute, recall, is this: error statisticians claim that it is not just the specific data produced that counts; we must know the testing plan that produced the data. Bayesians disagree, saying identical data should have identical implications.

Bayesians often argue that allowing stopping rules to count means making confirmation relative to the private intentions of the investigator. In some ways this claim is misleading, since what the stopping rule concerns is a testing procedure, not merely the psychological state of the investigator (as Mayo 1996 forcefully argues). Yet is there any inherent reason why the psychological state of the investigator could never be relevant to confirmation? Outside of presupposing the logic of science picture, I see no *a priori* reason why the psychological state of the investigator must be irrelevant. In actual science there are plenty of cases where the psychological state of the investigator is considered relevant evidence indeed. For example, many scientific journals now require investigators to disclose their financial interests in the research being reported. One obvious reason for making that information available is that it is relevant to evaluating the reliability of the research — because it tells us about potential bias. Double-blind studies are another case in point. If I ask whether a study was double blind, I am asking about the psychological state of the researchers, again doing so because it is relevant to evaluating reliability. Psychological states can be relevant.

The debate over stopping rules is a debate with strongly partisan positions and an acerbic history, and thus one with lots of baggage. The phrases 'stopping rules,' 'data that could have been gotten but wasn't', 'the error characteristics of the test,' and 'the likelihood principle' can have rather different connotations that are often buried in debate. To see how Bayes' theorem can take into consideration stopping rules and related notions, we need to stop first to make some distinctions.

Let's distinguish two different ways to describe a test — what I shall call the actualist and the logical description for lack of better names. The logical description refers to the traits that tests have on infinite repetitions. The actualist description picks out the characteristics of test in a circumscribed finite set of applications. An example of the latter is 'the best current seroconversion HIV test has an approximately 5% false positive rate.' This is the test's historical rate of giving positive results for those without the disease. 'The probability (long-run frequency) of the data 6 H/4 T from the procedure of flipping until 6H are reached is . . .' provides an example of a logical description.

Two different notions of stopping rules, of data we could have gotten, and of error characteristics come from this distinction. Logical stopping rules determine the outcome or sample space in a set of infinitely many hypothetical repetitions. An actualist stopping rule makes no reference to the infinite repetitions. Actualist descriptions allow us to talk about data we could have gotten but did not in actual cases. When an individual with HIV tests negative, then we could have with 5% probability gotten a positive. That result which we did not get in this case but could have can then be relevant to interpreting this particular test result. Similar differences show up for error characteristics. The HIV test has error characteristics. Those characteristics hold for specific finite populations. Significance tests also have error characteristics, but those are characteristics of infinite repetitions.

So are stopping rules and the rest compatible with Bayes' theorem? Actualist variants surely are. The widespread use of Bayes' theorem in interpreting diagnostic tests is a prime example. Prevalence of a disease in the population (the prior) and the sensitivity and specificity of the test (the likelihoods) are combined via Bayes' theorem to determine the probability of disease given a test result. The sensitivity and specificity of the test describe its error characteristics; they tell us what data we could have gotten that is relevant to our interpretation of the evidence. Actualist stopping rules are trivially relevant in measuring prevalence, specificity, and sensitivity because we must count cases divided by the total population observed.

Logical notions of stopping rules, etc. are not compatible with Bayes' theorem in that they provide us with no new information. I flip a coin and get 6H/4T. How often I would get that data in an infinite series of 'flip until you get 6H' is irrelevant since that is not what I did. Things would be completely different had I done a great many 'flip until you get 6H' trials and then reported the results of *one* such trial. That stopping rule and other possible data are informative.

Thus the way by which the data were generated, error characteristics, and related information can indeed be used by Bayesians. Bayesians have been so concerned to show the irrelevance of logical conceptions that they have denied or unduly downplayed the place that actualist notions can and must play. Of course it is once again an empirical issue just when stopping rules are informative.[6]

5. ERROR-STATISTICAL APPROACHES

I now turn to look explicitly at the error-statistical approach. The general moral I draw will be the same: rules of scientific inference are empirical in ways not fully appreciated.[7] The specific conclusions I reach are that (1) the error-statistical approach cannot avoid, as it claims, judgments about prior probabilities, (2) Bayesian criticisms of practices apparently endorsed by the error-statistical approach are compelling, and (3) Mayo's demand that any account of scientific inference must fit with statistical practice is thus misguided. My main approach shall be to look at two kinds of problems in standard statistical practice — the 'file drawer problem' and the 'base rate problem.' I argue that they establish both the general claim about the empirical nature of methodology and the specific ones about statistical practice. Before turning to those examples, however, I present and reject Mayo's arguments claiming to show that the error-statistical approach avoids the use of subjective prior probabilities, thus providing another argument that error-statistical approaches cannot avoid substantive empirical information.

Mayo claims for the error-statistical approach a kind of objectivity which Bayesian approaches cannot have. Error statisticians hold that 'what the results do and do not indicate is not a matter of subjective belief' (Mayo, 1996, p. 149); unlike the Bayesians, the error-statistical approach holds that 'knowledge of the world ... is best promoted by *excluding* so far as possible personal opinions' (Mayo, 1996, p. 82). Although she acknowledges that error-statistical methods must sometimes presuppose background knowledge, she denies that this involves any element of subjective personal opinion. Neyman-Pearson theory 'provides extensive methods for checking whether such bias is avoided'. (1983, p. 335). One such method is randomization, which Mayo clearly takes to ensure that a real effect is observed. (1996, p. 16; 1983, p. 335).

The error-statistical approach avoids other problems plaguing Bayesians according to Mayo. Bayesians face the problem of the 'catchall' hypotheses — roughly, competing hypotheses that say 'no factor I haven't thought of was present' — and error statisticians do not. Discussing Eddington's test of general relativity, Mayo denies that it required considering some ill-defined hypotheses about unthought of alternative theories of gravity (the catchall). Eddington's statistical methods did answer the question 'are the deflections on these plates due to gravity?' That question can be answered with reliability by error-statistical methods without considering the probability of a catchall about unknown theories of gravity, contra Earman (1993).

Bayesians, however, have no such out: they must calculate the value of the catchall, yet there is no objective way to do so.

These claims are not compelling. Mayo has an unwarranted confidence in the ability of statistical methods alone to rule out error (another instance of the logic of science ideal at work). Randomization, which she cites in this mode, is a prime example. It is common lore that randomization ensures that unknown factors are controlled or 'subtract out' as she puts it. But it is obvious — and many others have made this point before (see Urbach 1985) — that it does no such thing. One problem concerns the fact that any particular random assignment may well be unbalanced even if in an infinite sequence of randomizations balance would be expected. This is why practicing researchers often randomize and then reassign individuals if an imbalance is perceived. An even more serious problem is that randomization does nothing for confounding factors associated with the treatment. Randomize all you will to treatment and control groups, if a treatment has an unknown associated causal effect, that effect will not be subtracted out. For example, individuals were randomized to controls and treatment groups in studying the effect of oat bran on serum cholesterol. That randomization did nothing to rule out the possibility that the treatment, eating bran muffins in this case, was associated with consuming less fatty food, the true cause of the effect.

There is nothing special about randomization in this regard. Reliable use of statistical methods in general depends strongly on having the correct background beliefs. (By downplaying that dependence the error-statistical approach again shows its implicit sympathies for the logic of science ideal). Among the many such assumptions are beliefs that errors are independently distributed and claims about the functional nature (linear, quadratic, etc.) of the relationship. It is important to note that these are technical assumptions of statistical tools and do *not* include the extensive knowledge involved in solving the base rate and file drawer problems to be discussed below.

Once we recall how extensive the background knowledge must be in statistical testing, then the claim that Bayesians rely on subjective opinion while error statisticians report the objective probabilities of a test procedure looks bogus. It is helpful here to be clear about what we mean by 'subjective.' A subjective result might mean (1) a reported result that is dependent in some sense on someone's prior beliefs or (2) a reported result that is arbitrary, untestable, etc. Being subjective in the first sense need not entail being subjective in the second. If error statisticians report objective results, this surely cannot mean they report results that in no way depend on their beliefs. Unless we make very strong realist assumptions, error statisticians report what they *believe* to be the case about objective probabilities. Among the factors influencing those beliefs must be their beliefs about the relevant confounding variables and so on. Thus the error-statistical approach is subjective in the first sense of the term.

Of course it need not be subjective in the second sense, but then neither must Bayesian accounts. While Mayo makes it sound as if Bayesian inference gets its priors from out of nowhere, that is not the case. Presumably those beliefs result from earlier applications of Bayes' theorem plus background knowledge, and this is in the abstract no different than the error statistician who uses background knowledge resulting from previous statistical tests to conduct new ones.

Similar points can be made about the catchall problem. The problem may come up quite explicitly in Bayesian analyses, but the error-statistical approach faces the same problem. Consider the Eddington example. Eddington rejected the null hypothesis that the observed deviations were due to chance by standard statistical testing. Mayo repeatedly asserts that when we reject such nulls, we can assert that 'there is a genuine correlation' (1996, p. 300) or real effect. So Eddington's statistical analysis gave us good reason to think that the observed deflections of light were due to gravity, and yet, according to Mayo, Eddington had no reason to calculate the probability of a catchall about other unknown theories of gravity.

Could Eddington thus avoid considering the catchall? It depends on what we take the catchall to be. One great merit of Mayo's discussion of testing is the emphasis on the piecemeal or layered nature of testing. So Mayo seems correct that Eddington did not have to rule out the indefinitely many different theories of gravity mentioned by Earman, *so long as they do not countenance variables that might have confounded Eddington's findings.* But Eddington could not avoid the catchall of a second sort that says 'and there are no other unknown factors that could have systematically biased my data.' He no doubt believed his prior knowledge gave him good reason to doubt that such factors existed, but that knowledge is unavailable to the error-statistical approach which eschews such things in theory (but not in practice as we have seen). No procedure by itself, bereft of significant background knowledge, will the solve this catchall problem. That it could again reflects the vain hope for a logic of science.

These conclusions can be bolstered by noting the place that empirical information about the social organization of science must play in assessing error-statistical results. One phenomenon that shows the importance of social information is the file drawer problem. The file drawer problem is so called because it concerns the percentage of studies that are published compared to the total done. Journals frequently will not publish results that are not statistically significant. That means studies may be undertaken and then filed away if they show negative results. This practice, however, causes problems. If the negative results are all buried and Type I errors are possible, then the literature itself may give misleading indications of the truth (see Rosenthal 1979; 1993).

Consider for example a statistically significant result reported in a journal. Is that result well confirmed? For the error-statistical approach, we have to ask whether the result was produced by a severe test. Yet that question cannot be answered until we describe what the test was, and there are two ways to do that. Described narrowly, the procedure was significance testing by the researcher. Described widely, the test was for many researchers to perform significance tests and report the significant one. On the narrow description, this result is well confirmed and on the wide it is not. (This situation is not unique to the file drawer problem. Something quite similar occurs when a single researcher runs many regressions and then reports only the statistically significant ones.)

The conclusion to draw is that the error-statistical approach depends crucially on substantive empirical information, particularly social information. Looking at the error-statistical rules alone plus the data in the study will not determine whether the hypothesis at issue is well supported, contra the logic of science ideal. We need instead substantive empirical information — for example, how many file drawer studies are there? Moreover, the information we need essentially concerns the social structure of science. What are the policies of journals? How is information about unpublished studies disseminated? What are the incentives for doing repeated studies?

Aside from illustrating broader themes, the file drawer problem raises doubts about one standard of adequacy Mayo invokes in favor of her account and against the Bayesians. She claims that any philosophy of statistics must be consistent with the actual practice of science — with the use of statistics in the sciences (1996, p. xii, p. 362). However, Bayesians reject a great many standard statistical practices in science while the error-statistical view, she claims, is consistent with them. This is thus supposed to be strong evidence in favor of the error-statistical approach over the Bayesian. But that is doubtful, for the file drawer problem poses the following dilemma. Journals widely refuse to publish negative, i.e. non-significant, results. Data mining — using *try and try again* — is common as well. Much standard statistical practice seemingly violates error-statistical rules.[8] So Mayo cannot both advocate error-statistical standards and require accounts of inference to fit the actual statistical practice of scientists. Either those rules are mistaken, or they are not justified by their fit with standard statistical practice, and a major criticism of the Bayesian view falls away.

Now there is one reasonable reply here — argue that the naturalized epistemological justification was misguided all along. One way to do that would be a Bayesian one of taking the correct justification for rules of inference to depend on *a priori* truths of the probability calculus. Obviously that is a route the error-statistical approach wants to avoid.

Alternatively, we can argue that the naturalism project be differently conceived. Rather than simply fitting with the practice of science, the relevant criterion should be the reliabilist one of asking what is the effect of a given methodological rule on our actual empirical chances of getting to the truth. Though this approach is rejected by Mayo, nothing inherent in the error-statistical view requires it, and there are convincing reasons for thinking it plausible (see Laudan 1990; Kitcher 1993).

However, using such means-ends justification faces a different problem. It is not at all clear that error-statistical rules in actual fact do lead to a high ratio of truth even when properly applied, that is, when the file drawer or data mining problems do not exist. It is this claim I want to argue for next by discussing what I call the 'ignoring the base rate problem.'[9] It is a compelling illustration that the error-statistical approach cannot avoid empirical assumptions and prior probabilities without itself falling into error.

The error-statistical approach finds a hypothesis well supported if it has passed a severe test. A severe test is one with what Mayo calls (mistakenly I will suggest) a low error probability. A low error probability for her means that the odds are small that I will accept a hypothesis when the hypothesis is in fact false. In Bayesian terms, the likelihood *p(e/not h)* is low — we are unlikely to see a positive result when the hypothesis is false. It is not hard, however, to show that this notion of low error probability is misguided, because it is entirely compatible with rampant error. If I follow the rule 'take *h* as well confirmed when it passes a test with low error probability,' using Mayo's definition of the latter, it is entirely possible that I will believe false hypotheses with great regularity, depending on what the world is like.

One way to see the problem is to note that Mayo makes *p(e/not h)* central to deciding whether to believe *h* after it passes a test. But deciding whether to believe *not h*, given the test result *e*, is to make a judgment on *p(not h/e)*. But *p(not h/e)* cannot be calculated from *p(e/not h)*, and equating the two can easily lead to error, ie., to believing false hypotheses. That can happen when no attention is paid to the other factors that go into Bayes' theorem — *p(e/h)* and, above all, *p(h)* and *p(not h)*, which define what is often called the 'base rate.'

An example will help make this clear. A patient visits a doctor for a breast cancer test. The test turns up positive. Breast cancer occurs in .001 percent of the women who walk into the doctor's office for tests. The test is 100% sensitive, so if a patient has the disease, the test shows up positive. The test is 95% specific, so if the patient does not have the disease, 95% of the time the patient will show up negative. Suppose the test turns out positive. Should the doctor think she has good reason to believe that her patient has breast cancer? On the error-statistical view that question is answered by asking whether the test was severe and that requires a low probability of rejecting the null when it is in fact true. The following chart presents some possible numbers of cases realizing the above set of facts.

State/Test	Have Disease	No Disease
Positive	1	50
Negative	0	950

Table 1

Let *h*: the patient has breast cancer, *not h*: the patient does not have breast cancer, and *e*: test shows positive. Then *p(e/not h)* is .05 — reading down the right column, only 50 out of 1000 times when they don't have the disease will I see a positive and conclude they do (in other words, $p = .05$). So the test is seemingly severe. But given the numbers in Table 1, it is clear that if I practiced what the error severity rules preach, my own error probabilities would be quite high. Since I reject the null when a severe test gives me a positive, then the odds of making an error are represented by *p(not h/e)*. But by Bayes' theorem, that comes to:

$$p(\text{not } h = .999) \times p(e/\text{not } h = .05) / p(\text{not } h = .999) \times p(e/\text{not } h = .05) + p(h = .001) \times p(e/h = 1)$$

and thus the probability of error is 98%. These probabilities need not be taken as subjective probabilities; they are, we are imagining, real frequencies. Thus doctors who consistently follow the error-statistical strategy will err often.

There are several ways for error-statistical approach to respond. One strategy is to claim that this example is highly artificial and not at all typical of the way statistical inference is done. It is true that having objective numbers for the priors is atypical. Yet researchers are frequently in situations quite analogous to this example. Take, for example, a clinical trial of a new treatment that shows the treatment has a statistically significant effect and the test has high power. So like our previous example, the test is highly specific and sensitive. However, suppose, as is often the case, that researchers have no idea of the mechanisms of action of the new drug. Suppose, as is generally the case, that most positive results in clinical trials of this sort of agent have not panned out on further investigation. Then a researcher who gets the positive outcome (i.e. statistically significant) is in a very similar situation as that of the doctor interpreting the positive result on the cancer test. A reasonable researcher will remain skeptical despite the statistically significant result. This is in fact how clinical researchers reason all the time.[10]

A second strategy is to argue that the cancer test is not severe, for 'a failure to reject H with e ... does not indicate H so long as there are alternatives to H that would very often produce such a result. For example, the test might often yield a positive result even if only a benign type of breast condition exists' (Mayo, 1997, p. s202).

Note first that the example given here is *not* one where the alternative hypothesis — no breast cancer — entails the same evidence. The *p(e/not h)* is .05; the test does not often give a positive when no disease is present. So we cannot claim the test is not severe because it does not rule out competing explanations in this sense. Thus if the breast cancer test is not severe because of plausible competing explanations, its lack of severity will have to be because *p(not h)* is high, which it indeed is. But *p(not h)* is a prior, and the error-statistical approach claims to eschew priors. In fact, Mayo's phrase 'the test might often yield a positive result' cannot be given a definite meaning without specifying the base rate *p(not h)*; the frequency of false positives cannot be determined without it. Yet Mayo wants to say false positives are likely and thus the test not severe, but it is hard to see how to make that claim without knowing the priors.

There are further ways around this problem, but they raise other difficulties. It might be suggested that the above example seems a severe test only because we have misdescribed it. Perhaps we should describe the test here more richly as 'affirming statistically significant hypotheses about a population with a low base rate'. Then, this testing procedure will not be severe, because it does err frequently as we saw above. Since I have argued above that nothing in the Bayesian formalisms dictate how tests should be described, I should allow the same for the error-statistical approach.

However, this solution comes with a heavy price. First, we are explicitly building in priors, something the error-statistical view denies are relevant. So this is saving severity by giving it a Bayesian rationale. Second, the naturalized epistemological justification for the error-statistical approach is threatened. Following the rule 'take as well confirmed those hypotheses that pass significance tests' is the standard statistical practice in those sciences using statistical testing, *not* the revised rule discussed above that incorporates priors. So this leaves the error-statistical approach with a dilemma: either condemn the actual practice of statistics and thus give up the naturalized epistemological justification, or advocate a methodology that can be error prone.

Of course, it could be that as a contingent empirical matter most populations studied by statisticians have a base rate such that the error-statistical rules still produce good results. But I know of no reason to think that — for example, the breast cancer numbers cited earlier are not wildly unrealistic. Even if they were, this would be a minor victory for the error-statistical approach. Its value would, as I have been arguing for methodology in general, depend strongly on contingent empirical facts. It would also have a natural Bayesian diagnosis.

Finally, a third reply to the base rate problem is to claim that the base rate examples depend on Bayesian intuitions (Mayo 1997). Mayo takes the Bayesian to be committed to telling the patient who tests positive that they do not have the disease, since the posterior probability is so low. This is misleading. It is certainly true that the example allows us to tell the patient that in 98% of these positive test cases the patient does not have the disease. However, that is an objective fact, not an 'intuition' open to dispute. More importantly, after the test the probability of having the disease increases $p(h/e) > p(h)$ and for Bayesians like Howson and Urbach that is one measure of the extent to which the data *disconfirm* the hypothesis of no disease. Hence this example and its interpretation do not rely on skewed Bayesian intuitions.

So the base rate problem will not go away. That should be no surprise if we think of these issues in a larger context. Error-statistical approaches are a form of reliabilism, albeit an *a priori* one. But the debates over reliabilism suggest that there is no way to identify adequate notions of reliability that are not specified relative to the worlds in which they apply. This is just what we have seen in the error-statistical case. Error-statistical rules will work if the population is right, if the base rates are not as in the breast cancer example. Error-statistical rules will also work if we describe them in a way that specifies their domain of application — if we in effect build in the priors to the description of the test. These points seem just instances of the fact that reliabilist theories must be specified according to the world of application.

6. CONCLUDING CONTEXTUALIST MORALS

Thus my conclusions about the Bayesian/error-statistical dispute are as promised: The logic of science ideal causes both approaches to miss important ways that scientific inference is empirical and social. Recognizing these elements leads us to conclude that some standard Bayesian objections to error-statistical practices are misguided, for those practices are in fact compatible with the spirit of Bayesianism. And it also suggests that various standard error-statistical complaints again Bayesianism are likewise mistaken. Both views claim more for methodology than it can deliver.

These conclusions then provide some support for the general view of realism sketched at the beginning. Neither Bayes' theorem nor frequentist rules tell us the methodological traits that suffice to take a particular domain realistically. Both, properly understood, have their place in scientific argument. But they have to be embedded in general empirical arguments. The social sciences may use the machinery of significance testing, but contra Trout that does not tell us that such work is reliable, as the Bayesian criticisms of significance testing make clear. No doubt simplicity and the like are important elements in scientific reasoning as Maxwell claims, but as our discussion of Bayesianism shows, there is no interpreting and evaluating such criteria without substantive empirical knowledge. There is no purely philosophical argument employing those criteria that will establish realism. Scientific realism is a local issue.[11]

University of Alabama at Birmingham

REFERENCES

Bernardo, J. and Smith, A.: 1994, *Bayesian Theory*, John Wiley and Sons, New York.

Boyd, R.: 1990, 'Realism, Conventionality, and 'Realism About'', in G. Boolos (ed.), *Meaning and Method*. Cambridge University Press, Cambridge.

Campbell, R. and Vinci, T.: 1983, 'Novel Confirmation', *British Journal for the Philosophy of Science* 34, 315-341.

Carnap, R.: 1950, *Logical Foundations of Probability*, Routledge and Kegan Paul, London.

Crasnow, S.: 2000, 'How Natural Can Ontology Be?', *Philosophy of Science* 67, 114-132.

Day, T. and Kincaid, H.: 1994, 'Putting Inference to the Best Explanation in Its Place', *Synthese* 98, 271-295.

Earman, J.: 1992, *Bayes or Bust: A Critical Examination of Bayesian Confirmation Theory*, MIT Press, Cambridge.

Earman, J.: 1993, 'Underdetermination, Realism, and Reason', *Midwest Studies in Philosophy 18*, Edited by P. French, University of Notre Dame Press, Notre Dame.

Efron, B. and Tibshirani, R.: 1993, *An Introduction to the Bootstrap*, Chapman and Hall, New York.

Goldman, A.: 1987, 'Social Epistemics', *Synthese* 73, 109-104.

Harman, G. : 1965, 'Inference to the Best Explanation', *Philosophical Review* 74, 88-95.

Howson, C. and Franklin, A.: 1991, 'Maher, Mendeleev, and Bayesianism', *Philosophy of Science* 58, 574-586.

Howson, C. and Urbach, P.: 1993, *Scientific Reasoning: The Bayesian Approach,* Open Court, LaSalle, Ill.

Jones, R.: 1991, 'Realism about What?', *Philosophy of Science* 58,185-202.

Kincaid, H.: 1997, *Individualism and the Unity of Science: Essays on Reduction, Explanation and the Special Sciences*, Rowman and Littlefield: Lanham, MD.

Kincaid, H.: 2000, 'Global Arguments and Local Realism About the Social Sciences', *Philosophy of Science* 67, s667-s678.

Kitcher, P.: 1993, *The Advancement of Science*. Oxford University Press, Oxford.

Laudan, L.: 1990, 'Normative Naturalism', *Philosophy of Science* 57, 60-78.

Maher, P.: 1988, 'Accommodation, Prediction, and the Logic of Discovery', *PSA 1988,* edited by A. Fine and J. Leplin, Philosophy of Science Association, East Lansing, 381-392.

Maxwell, G.: 1970, 'Theories, Perception and Structural Realism', in R. Colodny (ed.), *The Nature and Function of Scientific Theories,* University of Pittsburgh Press, Pittsburgh.

Mayo, D.: 1983, 'An Objective Theory of Statistical Testing', *Synthese* 57, 297-340.

Mayo, D.: 1996, *Error and the Growth of Experimental Knowledge*, The University of Chicago Press, Chicago.

Mayo, D.: 1997, 'Error Statistics and Learning from Error: Making a Virtue of Necessity', *Philosophy of Science* 64, s195-s212.

McCloskey, D. and Ziliak, S.: 1996, 'The Standard Error of Regression', *Journal of Economic Literature* 34, 97-115.

Motulsky, H.: 1995, *Intuitive Biostatistics*, Oxford University Press, Oxford.

Rosenthal, R.: 1979, 'The File Drawer Problem and Tolerance for Null Results', *Psychological Bulletin* 86, 638-641.

Rosenthal, R.: 1993, 'Cumulating Evidence'. In G. Keren and C. Lewis (eds.), *A Handbook for Data Analysis in the Behavioral Sciences: Methodological Issues*, Erlbaum, Hilsdale, NJ, 519-559.

Royall, R.: 1997, *Statistical Evidence: A Likelihood Paradigm*, Chapman and Hall: New York.

Trout, J. D.: 1998, *Measuring the Intentional World: Realism, Naturalism, and Quantitative Methods in the Behavioral Sciences*. Oxford University Press, Oxford.

Urbach, P.: 1985, 'Randomization and the Design of Experiment', *Philosophy of Science* 52, 256-273.

van Fraassen, B.:1980, *The Scientific Image*, Oxford University Press, Oxford.

Wessels, L.: 1993, 'Scientific Realism and Quantum Mechanical Realism', *Midwest Studies in Philosophy* XVIII, 317-332.

Williams, M.: 1996, *Unnatural Doubts: Epistemological Realism and the Basis of Scepticism*, Princeton University Press, Princeton.

NOTES

[1]The general contextualist idea takes its inspiration from Michael Williams' (1996) attempt to show that skepticism and other epistemological problems presuppose unwarranted essentialist notions. In philosophy of science it is of a piece with various attempts to show that realism cannot usefully be treated as a global question but is more fruitfully assessed as specific questions about specific entities with specific domains (Wessels 1993; Jones 1991; Kincaid 2000; Crasnow 2000).

[2]The conception of methodology as a logic of science is of a piece with the idea that the social structures of science — authority relations, the intellectual division of labor, and so on — are irrelevant to assessing scientific inference and methodology. I shall defend the opposite view: to assess whether a given piece of evidence supports an hypothesis often requires reference to social information about the practice of science. If such information can be relevant, then methodology would certainly not be entirely *a priori*. Nor would it be universal or formal, for determining good inference practices would rely on contingent and particular facts about how science is organized. I shall argue that such possibilities are indeed real for evaluations of statistical practice and thus provide further evidence for my claim that neither Bayesians nor error statisticians provide anything close to a logic of scientific inference. The idea that information about the social structure of science is essential to confirmation may seem highly unorthodox. Yet it has some strong motivations. One rationale comes from the reliabilist tradition in epistemology. That tradition determines justification

according to whether belief-forming mechanisms promote the truth. But it is an easy step from that question to further ones about the influence of social processes on belief (see Goldman 1987; Kincaid 1997). A second and related motivation comes from looking at the actual practice of science. Deference to authority and trust are integral to the specialization that dominates modern science. If actual scientific practice is to ever meet philosophers' standards of confirmation, those standards will have to make room for the social practices that ground authority and the like.

[3]Bayesian and error-statistical approaches do not of course exhaust the possible views, with likelihood inference (Royall 1997) being a particularly salient third possibility. It is beyond the scope of this paper to examine the extent to which the arguments and objections that follow bare on likelihood approaches except where they share claims with the Bayesians, e.g. on the irrelevance of stopping rules.

[4]See Howson and Urbach (1993, pp. 210-215), for examples.

[5]There is another way Bayesian approach may have to use empirical information — they may be required to factor in information about the social processes of science. The previous remarks about conflict of interest statements and double blind studies suggest why. To the extent that authority relations, trust, and the like are an inevitable part of scientific practice, rational agents updating their beliefs will have to take them into account. There are a multitude of ways such assessments can be incorporated in applying Bayes' theorem. Certainly the prior plausibility of hypotheses may be influenced by who advances them. Their advocates may have great past track records, may be known for high quality or sloppy work, and the like. Relations of authority and trust are particularly salient when it comes to assessing the reliability of evidence. In Bayesian terms, this shows up in the probability assigned to the alternative hypotheses that observed data are due to errors of various kinds.

[6]Critics of Bayesianism argue that because Bayesians ignore stopping rules, it is possible to reason to a foregone conclusion. There are several replies to this charge. First, as I have argued they need not ignore all stopping rules. The logical stopping rules involved in significance testing they do ignore as uninformative. This allows for the possibility of repeated significance testing leading to a foregone conclusion if statistical significance is used as a decision criterion. But that is not the criterion Bayesians advocate, since it leads to violations of Bayes theorem (see Bernardo and Smith 1994).

[7]There is one obvious way in which the error-statistical approach is committed to *a priori* ideals. It argues on conceptual grounds that the only notion of probability that makes sense is that of objective frequency, thereby deciding on *a priori* grounds how scientific inference must be understood. But it is also committed in that it claims that it provides 'ampliative rules for generating, analyzing, and learning from data in a reliable and intersubjective manner' (Mayo, p. 59), all ultimately based on one rule, that of the severe test. Mayo rejects inductive logic versions of the logic of science ideal because she rejects the ideal of a degree of confirmation, but that is not rejecting the logic of science ideal in the broad sense that I am using the term.

[8]For an attempt to assess how widespread similar problems are in economics, see McCloskey and Ziliak (1996).

[9]I am not the first to note this problem — it is common both in the philosophical literature and in texts on, for example, clinical decision making.

[10]See Motulsky (1995) for example.

[11]Thanks to Sara Vollmer, Deborah Mayo and an anonymous referee for detailed and helpful comments on earlier drafts, even though neither is convinced by major parts of the argument.

TIMOTHY D. LYONS

SCIENTIFIC REALISM AND THE PESSIMISTIC META-MODUS TOLLENS

1. INTRODUCTION

Broadly speaking, the contemporary scientific realist is concerned to justify belief in what we might call *theoretical truth*, which includes truth based on ampliative inference and truth about unobservables.[1] Many, if not most, contemporary realists say scientific realism should be treated as 'an overarching scientific hypothesis' (Putnam 1978, p. 18). In its most basic form, the realist hypothesis states that theories enjoying general predictive success are true. This hypothesis becomes a hypothesis to be tested. To justify our belief in the realist hypothesis, realists commonly put forward an argument known as the 'no-miracles argument'. With respect to the basic hypothesis this argument can be stated as follows: it would be a miracle were our theories as successful as they are, were they not true; the only possible explanation for the general predictive success of our scientific theories is that they are true.[2]

In light of historical concerns, realists have modified the basic position, changing their focus from general predictive success to novel success, from truth to approximate truth, etc. In this paper, I will challenge a number of variations on this explanationist argument and the hypothesis it is meant to justify. I will clarify that the threatening historical argument is not a 'meta-induction' from which we infer the falsity of contemporary theories; rather, it is a valid deduction that bears on the justification of our beliefs. After articulating the implications this argument has for realism, I will show that this argument poses a serious threat even when success is understood as novel success. I will also challenge the claim that the approximate truth of a theory can explain its success and will offer an alternative non-realist explanation. My approach will be to work my way from the naive formulat-

S. Clarke and T.D. Lyons (eds.), Recent Themes in the Philosophy of Science, 63–90.

-ion of realism toward more sophisticated versions in light of these non-realist objections.

2. LAUDAN'S CONFUTATION

With the basic realist position spelled out above, we can turn to an important historical argument against realism. Larry Laudan, in his 'Confutation of Convergent Realism', provides a frequently cited list of theories and existential postulates that were successful but which are, according to contemporary theories, false (1981, pp. 121-122):

— the crystalline spheres of ancient and medieval astronomy;
— the humoral theory of medicine;
— the effluvial theory of static electricity;
— 'catastrophist' geology, with its commitment to a universal (Noachian) deluge;
— the phlogiston theory of chemistry;
— the caloric theory of heat;
— the vibratory theory of heat;
— the vital force theories of physiology;
— the electromagnetic aether;
— the optical aether;
— the theory of circular inertia;
— theories of spontaneous generation.

Realists have been responding to Laudan's 'Confutation' for two decades now. One question that has become salient: what sort of inference is being drawn from this list? Many realists (I dare say nearly all) assert that the article containing this list provides an exemplary articulation of 'the pessimistic meta-induction'. In regard to Laudan's 'classic paper "A Confutation of Convergent Realism"', Jarret Leplin tells us Laudan 'argues *inductively* that our current picture is *unlikely*, in retrospect, to have achieved referential success or approximate truth from the vantage point of future science, and that realism, accordingly, has no application' (1997, p. 137) [my italics]. Herman De Regt interprets Laudan's argument in a similar way, 'if so many successful theories turn out to be ontologically false we may *induce* the conclusion that current successful theories will *turn out to be* ontologically false as well' (1994, p. 11) [my italics]. 'Laudan's "historical gambit"', says Stathis Psillos, '*concludes, inductively*, that *any* arbitrarily successful scientific theory Tn+1 is *likely to be false* (or at any rate, *more likely to be false* than true)' (1999, p. 105) [my italics]. Each of these philosophers is taking Laudan's Confutation paper as an articulation of the pessimistic *meta-induction* toward the conclusion that *our present theories are probably false*.

However, I wish to suggest that this very common reading constitutes a significant misinterpretation of Laudan's argument. First, and briefly, contra the quotes above, Laudan's list is not presented as evidence that our contemporary theories are probably false (or even that it is *unlikely* that we have reached the truth). He is concerned with the justification of our beliefs, not drawing conclusions regarding the truth value of our contemporary theories.[3] A second and more important point: against each of the quotes above, I submit that Laudan is not arguing inductively at all.[4] Before outlining the proper understanding of his argument, we can note that construing Laudan's argument as an induction gives the impression that his inference is insufficiently supported. In specific reference to Laudan's 'Confutation', Psillos writes, 'This kind of argument can be challenged by observing that the inductive basis is not big and representative enough to warrant the pessimistic conclusion' (1999, p. 105).[5] Additionally, we can note that the intuition that drives us to consider many inductions credible may well not ground an induction of this sort. While we may arguably be justified in assuming, for instance, that there is a natural uniformity in respect to some *physical* entities and their properties, such a uniformity need by no means carry over into the realm of scientific theories. We have been given no reason to think there would be a uniformity of overall falsity here.[6]

However, I contend that Laudan is making no induction and that such criticisms are therefore misdirected. His list of historical theories that were successful but false is presented (and ought to be presented) as phenomena which serve to falsify the epistemic realist's hypothesis (and to challenge the explanationist argument that grounds that hypothesis). Any argument of this form is quite properly distinguished from a deductively invalid induction. It concludes with no universal claim, nor even a prediction. No inference is drawn about the future, or even the present. The list serves instead as fuel for the second premise of a logically valid *modus tollens* argument.[7] Again, in its most basic form, the hypothesis the realist says we can be justified in believing states that theories enjoying general predictive success are true. The *modus tollens* argument against that hypothesis reads as follows:

Premise 1: If the realist hypothesis is correct (A), then each successful theory will be true (B)

Premise 2: We have a list of successful theories that are not true (Not B)

Conclusion: Therefore, the realist hypothesis is false (Not A)

And, given the argument, we are not justified in believing the hypothesis, etc. (Further specific implications will be spelled out below.)

Even viewing the list as occupying the second premise of a *modus tollens*, one may be tempted toward two related contentions against that premise. First, though the non-realist is questioning whether we can justifiably believe that our theories are true, the non-realist appears to be assuming the privilege of *knowing* that these past theories are false. But, given, say, the Duhem-Quine problem, what would license the non-realist's claim to have such knowledge? Second, to say that past theories are false, the non-realist has to assume present day theories are true; this is surely unacceptable given that the non-realist is ultimately trying to show we have no justification for such a belief.[8] However, in employing the second premise of *the modus tollens*, the non-realist is not *required* to believe any such claims. Since successful present science tells us that much of successful past science is false, successful present science and successful past science cannot both be true. Therefore, we have a subset of theories that are successful yet false.[9] The only claim being made, in the end, is that there are some successful yet false theories in the set considered, and acknowledging this necessitates no specification of where the falsity resides. For ease of discussion, in what follows, I will nonetheless refer to *past* theories as being 'false', as being 'counterinstances', etc. But the richer understanding of such phrasings should be kept in mind.

3. NAIVE REALISM AND THE IMPLICATIONS OF THE META-MODUS TOLLENS.

Let us determine the effect Laudan's argument has against the naive version of the realist's explanationist argument introduced above. Actually, because our present concern is with naive realism, we do not need to limit ourselves to Laudan's list. Any theories — for instance, the theories of Copernicus, Galileo, Kepler, Newton, Lavoisier, Darwin, Mendel, Wegener, Bohr, etc. — that were successful but are now taken as 'strictly speaking, false' are relevant candidates. That noted, looking at history, we see many such theories. The following, more specific, points make clear the extent to which naive realism is threatened by the list of counterinstances. (I will draw the same implications regarding the more sophisticated realisms discussed below. Doing so, I will direct the reader back to this section. I ask the reader, when returning here, to replace the naive versions of the realist hypothesis and explanationist argument with the hypothesis and argument at issue in the relevant section. The same implications hold).

Take first the explanationist argument the realist employs to justify belief in her hypothesis. For naive realism, the key premise states, 'It would be a miracle were a theory successful but not true'. Given that there have been successful theories that are false, we must either take this key premise of the realist's inference to be absurd and therefore unacceptable, or we must hold the success of these false theories to be miraculous. Since no-one in this debate wants to say the list of concern is a list of miracles, the key premise must be rejected. A similar but separable point: the list constitutes a list of historical theories whose success the realist cannot explain (of the implications I'm drawing, this is one that Laudan himself emphasises). How then can naive realism constitute a good or even adequate explanation of the success of scientific theories in general? Realists say non-realism is to be rejected for the very sin that realism itself is here committing, failing to explain success. Further, since we are rejecting miracles, and since realism cannot explain these successes, some explanation other than the truth of these theories is required to do so. And once that needed explanation is introduced (whatever it may be), the claim that truth is the only explanation for success is rendered futile. Moreover since that explanation would have to be acceptable for explaining the success of those theories on the list, it would also have to be acceptable for explaining the same sort of success of all other theories, including those of contemporary science. Thus it would have greater breadth than the realist's 'truth explanation'. Given that truth could at best only explain the success of some theories, and given that another explanation will be needed to explain the success of others, the truth of our theories would appear to be neither a good explanation nor the only explanation for the success of scientific theories in general. The argument by which the realist seeks to justify her hypothesis is rendered unacceptable.

Continuing, let us directly address the specific hypothesis the naive realist wishes to say we are justified in believing. For naive realism that hypothesis reads, 'successful theories are true'. Testing this correlation, we see first that, without begging the very question being addressed, i.e., without simply asserting the realist hypothesis to hold, we can establish not a single confirming instance of the hypothesis, not a single instance in which both success and theoretical truth obtain. Thus the hypothesis attains no solid inductive support whatsoever. Further, having now interpreted this historical argument as offering a set of counterinstances, in the proper spirit of Laudan's paper, we ask with Martin Carrier, 'What could count as empirical counter-evidence' for the realist hypothesis (1991, p. 25). And we answer: successful theories that are false. Given the high number of such theories, the hypothesis that successful theories are true has an exceptionally high number of falsifications. With the counterinstances in hand, we have a non-universal falsifying hypothesis that contradicts the realist hypothesis:

theories can be successful without being true. The high number of counter-instances provide significant empirical confirmation (or corroboration) to this hypothesis which directly falsifies the unconfirmed realist hypothesis. Testing the realist hypothesis, we discern not a single genuine confirmation and numerous falsifying instances. Thus, we cannot even say the realist hypothesis *probably* holds; even a mild Popperian *conjecture* that there is a positive correlation between truth and success would have to be rejected. (One is tempted to suggest that the realist hypothesis shares the evidential status of the hypothesis that all ravens are white.) We are prompted to ask ourselves whether any scientific hypothesis would be retained with such a weak track record. The answer appears to be, no. If we are to be sincere about testing our ideas against the world, then we need to take this evidence seriously: the evidence strongly suggests we must reject the realist's hypothesis as false. And, again, this is no induction: no universal generalization, nor even a prediction, has been made. As we address other realist formulations, we will be led to these same general conclusions against the realist inference and hypothesis. I will refer back to these conclusions collectively as *the implications of the meta-modus tollens*.

As I've implied, naive realism is a straw man. In response, the realist will here point out that we are not and ought not be *naive falsificationists*: we should be afforded the opportunity to modify our hypothesis. For example, our naive realist might deny that the tacit universality in the realist hypothesis is required. We can simply conjoin to our realist hypothesis the phrase, 'usually' or 'typically': successful theories are typically true.[10] A similar response would be to modify our hypothesis to say that a certain percentage of our successful theories are true.[11] However, these modifications do not get us very far — even putting their suspiciously *ad hoc* character aside. Without begging the question of realism, these modified hypotheses have no positive, confirming instances, thus no standard inductive evidence. Moreover, they fail to explain the counter-instances. Finally, and perhaps most importantly, these modifications clash with the very argument the realist is using to support them, the no-miracles argument. For that argument says it would be a miracle if a theory were successful but not true. The tacit supplementary claim is that we ought not accept miracles. Yet the phrase 'typically' and the specification that a certain percentage of successful theories are true entail the possibility of theories that are successful but are not true; thus these modifications leave us conceding to the existence of miracles in science. So even ignoring their *ad hoc* nature, the meta-modus tollens cannot be averted by weakening the naive realist hypothesis to say that successful theories are typically true (see Brown on this (1985)) or that a certain percentage of them are true.

4. NEW REALISM: NOVEL SUCCESS AND TRUTH

We have thus far been concerned with general predictive success. Primarily in response to Laudan's list,[12] a new version of the realist argument is appealed to by many, one which focuses not on mere *predictive* success but on the *novel* success of scientific theories (Musgrave (1985, 1988); Lipton (1993, 1994); Psillos (1999); and Sankey (2001)). Alan Musgrave writes, 'The fact to be explained [by realism] is the (novel) predictive success of science' (1988, p. 239). Novel success can be divided into two sorts: temporal novelty, in which the phenomena predicted are unknown to scientists at the time the prediction is derived; and use-novelty, a less demanding form, in which the phenomena may be known but were not used in formulating the theory. For our present purposes, then, a novel prediction should be understood as a prediction of phenomena that were not used in the formulation of the theory, and either sort noted will qualify. The following are examples of novel successes favoured by realists:

General Relativity

- light bends around massive objects (confirmed by the Eddington expedition)
- the gravitational redshift of spectral lines

Special Relativity

- time dilation (confirmed by jets carrying atomic clocks)

The appeal to novel rather than predictive success brings an element of wonder that ostensibly gives new force to the realist's no-miracles argument. It is extremely unlikely — in fact, says the realist, it would be a miracle — if a false theory would lead us to 'hitherto unknown' phenomena, or phenomena that were never *used* in the formulation of our theory. Novel success, it is held by the realist, has obvious epistemic significance. Further, redefining *success* in this way appears to hold the promise of eliminating many if not all of the theories in Laudan's list. Musgrave claims in response to Laudan's list that 'few, arguably none, of the theories cited had any *novel* predictive success' (1985, p. 211, footnote 10). If the theories on Laudan's list enjoyed *only* general predictive success, they do not qualify as genuinely successful, and the realist hypothesis is not falsified.

At this level the realist is trying to justify our belief in the following hypothesis: theories enjoying novel predictive success are true. The realist justification for this claim is that the truth of the theory is the only explanation for its novel success; it would be a miracle were a scientific theory to enjoy novel success were it not true. In his 'Confutation' article, Laudan points out that realists 'say little about what ... success amounts to'

(1981, p. 23). Presumably for this reason, Laudan was not concerned to present a list of novel successes. Testing the new realist package, then, calls for new data. We ask, in the history of science, are there novel successes from theories that have nonetheless turned out to be false? Attempting to begin the relevant list, I find there have been numerous examples such as the following:

Caloric Theory

- the rate of expansion is the same for all gases — confirmed by Dalton and Gay Lussac;[13]
- the speed of sound in air —predicted by Laplace;[14]
- the depression of freezing point by pressure — predicted by J. Thomson, using the Carnot theory, confirmed by W. Thomson;[15]
- steam engines with higher pressure have higher efficiency — predicted by Carnot;[16]
- the adiabatic law — derived by Poisson.[17]

Phlogiston Theory

- sulfuric acid and phlogiston (emitted from glowing coal) combine to create sulfur — predicted/confirmed by Stahl;[18]
- heating a calx with inflammable air (hydrogen) turns it into a metal — predicted/confirmed by Priestley.[19]

W.J.M. Rankine's 19th Century Vortex Theory

- the specific heat for saturated steam is negative (prior to Clausius's prediction of the same phenomenon);[20]
- the value of the specific heat of air under constant pressure (this conflicted with the data then available, yet was later confirmed);[21]
- the existence of the entropy function (predicted well before Clausius's formulation of the principle of entropy increase);[22]
- the cooling effect for carbon dioxide as later found in the Joule-Thomson expansion experiment.[23]

Newtonian Mechanics

- the existence of Neptune;
- the return of Halley's comet;[24]
- the oblate shape of the earth;[25]
- the non-Keplerian perturbations of planets;[26]
- the ability to use the moon in propelling a rocket back to earth — confirmed by NASA, Apollo 13.

Fermat's Principle of Least Time

- light slows in a dense medium — confirmed by Fizeau and Foucault.[27]

Fresnel's Wave Theory of Light and Theory of the Optical Ether

- the 'white spot': directing a point-beam of light at a small opaque disc will result in the appearance of a bright spot in the disc's shadow — predicted by Poisson, confirmed by Arago;[28]
- the 'black spot': directing point-beam of light at a circular hole in an opaque screen will result in the appearance of a black spot located behind the screen along the line of the beam — suggested by Poisson, derived and confirmed by Fresnel;[29]
- Numerous quantitative details regarding (previously known) straightedge diffraction cases, predicted and confirmed by Fresnel;[30]
- the existence of internal conical refraction;
- the existence of external conical refraction — both predicted by Hamilton, confirmed by Lloyd.[31]

Maxwell's Ether Theory

- the existence of radio waves — confirmed by Hertz.[32]

Dalton's Atomic Theory

- the varying weights of oxalate acid (and of carbonic acid, and of sulfuric acid) that react with a given weight of potash to make different compounds are in simple numerical relation — confirmed by Thomson and Wollaston; 33
- the same relation holds for the oxides of nitrogen and ethene and methane — confirmed by Dalton himself (all other post-1803 confirming instances of Dalton's Principle of Simplicity should be added here as well);[34]
- the individual constituents in a mixture of gases exert the same pressure as they would alone (Dalton's law of partial pressure);[35]
- the atomic compositions of carbon monoxide, carbon dioxide, nitrous oxide, nitric oxide, and nitrogen dioxide;[36]
- carbonic acid has a linear compound atom (ie., is a linear molecule) and sulfuric acid has a triangular compound atom (ie., is a triangular molecule).[37]

Kekulé's Theory of the Benzene Molecule

- the substitution of one hydrogen atom with five chorine atoms in Benzene results in one (rather than two) isomers — confirmed by Landenburg;[38]
- the number of isomers of the derivatives that come about by substituting one hydrogen, two hydrogens . . . six hydrogens — confirmed by Beilsten and Kurbatow;[39]
- benzene displays the properties of having three carbon-carbon double bonds — confirmed by J.W. Bruhl;[40]
- the ozonisation of oxylene (a benzene derivative) followed by hydrolysis yields glyoxal, methyglyoxal, and dimethyglyoxal — confirmed by Levine and Cole.[41]

Mendeleev's Periodic Law

- the existence of gallium — confirmed by de Boisbaudran;[42]
- the existence of germanium[43]— confirmed by Nilson;[44]
- the existence of scandium — confirmed by Winkler;[45]
- the existence of technetium — created by Perrier and Segre;[46]
- the existence of rhenium — confirmed by Noddack, Tacke, and Berg;[47]

- the existence of francium — confirmed by Perry;[48]
- the existence of plutonium — confirmed by Marie Curie;[49]
- beryllium has an atomic weight of 9 rather than 14;[50]
- uranium has an atomic weight of 240 rather than 120;[51]
- gold has a higher atomic weight than platinum, iridium, and osmium;
- platinum has a higher atomic weight than iridium and osmium;
- iridium has a higher atomic weight than osmium.[52]

Bohr's 1913 Theory of the Atom

- three series of the hydrogen line emission spectrum (in addition to the Balmer and Paschen series) — confirmed by Lyman, Brakett, Pfund.[53]

Dirac's Relativistic Wave Equation

- the existence of the positron — confirmed by Anderson and Blackett.[54]

The Original (pre-inflationary) Big Bang Theory

- the existence of the cosmic background radiation — predicted by Gamow, confirmed by Penzias and Wilson.[55]

It is important to emphasise that I've restricted my list to the most rigidly demanding sort of novel success, temporal novelty. We nonetheless have a list of novel predictive successes, all coming from theories that are false by present lights. (I will add to this list below.) According to contemporary science, heat is not a material fluid, and is not conserved; no substance is emitted from a flame; Newtonian mechanics holds neither at high speeds nor small scales; gravity is not a force; time is not absolute; mass is not independent of energy; there is no aetherial medium through which light as an elastic vibration travels; elemental properties are not a function of their atomic weights; the principle of least time has been replaced by the principle of least action; electrons do not have circular orbits, etc. Against the move to novelty, we have in hand a substantial new list of counterinstances. Adding perhaps the implication that novel success is not as special as the realist considers it, testing this form of realism, we are again led to the implications of the *meta-modus tollens* discussed in Section Three.

5. SOPHISTICATED REALISM: APPROXIMATE TRUTH AND NOVEL SUCCESS

The natural response (and the response the reader has probably been anticipating) is to concede that our appeal to truth is too bold and to make an appeal to approximate truth. This sophisticated realist package now asserts the hypothesis that theories achieving novel success are at least approximately true. And the argument meant to justify our belief in that hypothesis now reads, *it would be a miracle were our theories to achieve novel success were they not at least approximately true*. Before concerning myself with the historical counterinstances against this version of the package, I will challenge the sophisticated realist position from another, though no less significant, direction.

On the Likelihood of Success given that a Theory is Approximately True

Let us clarify the realist's situation. The realist wants to say those theories in the list are approximately true. These are theories that have not only been replaced by later theories, they have also, by present lights, encountered empirical failure. The latter point imposes a requirement on our notion of approximate truth. It must be sufficiently nonrestrictive to allow that an approximately true theory can significantly fail empirically. The desire to accommodate the list therefore mandates that our notion of approximate truth be quite a *permissive* notion. However, our realist is pulled in another direction as well. She says that the predictive success of science can only be *explained* by the approximate truth of our theories. This of course requires that approximate truth can explain success in the first place. While realists will likely include a number of additional premises, they employ a Peircean (abductive) argument whose fundamental claims and structure are as follows:

Premise 1: If a theory is approximately true, then it will (at least be likely to) enjoy novel success[56]

Premise 2: We have had theories that have enjoyed novel success

---

Conclusion: Those theories are approximately true

Given the centrality of the first premise, the realist's explanationist argument demands that our notion of approximate truth be quite *strict*. In short, realists have to make approximate truth permissive enough so that an approximately true theory can fail empirically; yet they must make it strict enough so that such a theory will be likely to be successful. That there is a tension between these requirements should be apparent and become more so.

Is premise 1 acceptable? It is based on the assumption that theories that approximate one another make the same (or approximately the same) empirical predictions. Laudan charges realists with failing to show that an approximately true theory will be successful (1981, pp. 29-32).[57] I wish to argue toward the stronger conclusion that the approximate truth of a theory *will not*, in itself, make success likely. First, a slight change in claims about unobservables can lead to dramatically different empirical predictions. Let us consider the most basic and non-contentious sort of approximate truth. Say an experimenter gets a value within one one-thousandth of the *actual* value. The realist (and many others) will surely want to say such a result would be approximately true. (In agreement with Brown (1985), I'd suggest it should be a condition of adequacy that a notion of approximate truth or verisimilitude include such scenarios.) Consider a set of theory complexes, C1-Cn that approximate one another in this fundamental way. C1 is the entire corpus of contemporary science. C2-Cn are identical to C1, except

- in C2, the charge of the electron is higher by one one-thousandth of its value in C1;
- in C3, the charge of the electron is higher by one one-billion-billionth of its value in C1;
- in C4, the charge of the electron is *lower* by one one-thousandth of its value in C1;
- in C5, the charge of the electron is lower by one one-billion-billionth of its value in C1;
- in C6, the charge of the *proton* is higher by one one-thousandth of its value in C1;
- in C7, the charge of the proton is higher by one one-billion-billionth of its value in C1;
- in C8, the charge of the proton is *lower* by one one-thousandth of its value in C1;
- in C9, the charge of the proton is lower by one one-billion-billionth of its value in C1;
- in Cn, . . .

We have a set of theory complexes all well within a range of being approximately the same. Indefinitely many additional complexes could be generated within the range of approximation delimited here. C1 enjoys an amazing and unprecedented degree of empirical success. However, despite the fact that each of the many alternatives C2-Cn approximates C1, each of these predicts that all matter would *repel* all matter. While C1 enjoys amazing empirical success, the approximating theories predict no universe at all. Therefore, *all theory complexes in our set of complexes approximating C1 completely fail empirically.*[58] I've used this example to illustrate that a slight change in claims about unobservables can — and in this case, does — lead to dramatically different empirical predictions.[59] Imagine for a moment that every statement in C1 is true. Despite the fact that C2-Cn would be approximating the truth (to an amazing degree!), no members of that class would enjoy any degree of empirical success, let alone that enjoyed by C1.

A second point toward the claim that the approximate truth of a theory will not, in itself, make success likely: the mere stipulation that a theory is approximately true makes no restriction on auxiliary hypotheses. If no auxiliaries connect the theory to the world, the theory need not enjoy empirical success. Assume for a moment that Leibniz's theory of monads is approximately true. In the context of contemporary science, this theory would not be conjoined to any auxiliaries that connect it to the empirical world. Thus, despite its approximate truth, it would not be successful. And even if the realist were to conjoin to her explanation the claim that the theory is connected to auxiliary statements, any auxiliaries whatsoever are allowed, from the irrelevant to the absurdly false. That given, an unlimited number of theory complexes can be generated, far more of which will fail than succeed. Thus, with no restriction on the nature of the auxiliaries, our approximately true theory is far more likely to fail empirically than to succeed. (This holds, as well, for the mere stipulation that our theory is true, but I will not pursue this issue here.)

The only hope the realist might have is to modify her position by adding the claim that the auxiliaries are, like the theory itself, approximately true. This restriction does not come without consequence. It makes the realist hypothesis bolder and more demanding. It thus decreases the potential approximate truth realism has for accommodating the list. (The tension I've noted above is at play.) Most importantly, this necessary restriction is not sufficient to solve the explanatory problem. Assume for illustrative purposes that we have a theory complex consisting of a true theory conjoined to a full set of true auxiliary claims. We then consider every possible modification of our theory, T1-Tn, within the range of being approximately true, while keeping the auxiliaries as they are. We've seen the results of the electron/proton example above. In the realist's favour, however, let us stipulate that, for some reason, in this case, modifying only our theory we are not led to such a dramatic situation. Nonetheless, a theory that is approximately true is false; false theories have an infinite number of false consequences; and, (to accommodate the list) our notion of approximate truth must be sufficiently broad that it allows for a great deal of empirical failure. Even though the resultant complexes contain a full set of true auxiliary statements, it is quite likely that a number of the complexes will encounter empirical failure.

Now take just one auxiliary constituent, A1. Consider the potentially numerous modifications of A1 within the range of being approximately true, A1a-A1n. We then consider every possible T + A1 conjunction that can be brought about from A1a-A1n and T1-Tn. The quantity of complexes generated would inflate by multitudes. Even holding all other auxiliaries true, we will likely have a great many complexes that fail. Take a second auxiliary, A2, and all modifications permitted within the range of maintaining its approximate truth, A2a-A2n. Adding each of these possibilities to the legion of T+A1 theory complexes, we have another expansive increase in the set of complexes. We imagine continuing this process, modifying each member of the full set of true auxiliary statements in our initial complex. Three factors will serve to proliferate the number of resultant complexes:

- the number of auxiliary statements included in the original complex;
- the number of ways each given statement can be modified so that it fits into our (necessarily permissive) notion of being approximately true;
- the number of possible conjunctions of all the statements that are available.

All possible modifications of the true statements in the initial complex and all possible conjunctions of these statements render together an innumerable mass of complexes whose statements are all approximately true. And the greater the number of possible complexes available, the greater the quantity of empirical predictions made by the set of all complexes. Among the set of complexes, there will be indefinitely many that will be empirically *unsuccessful*. Most importantly, failure is not restrictive; the ways in which, and degrees to which, a complex can fail to be empirically successful are not limited. By contrast, success is restrictive: the ways in which, and degrees to which, a complex can be successful are very restricted. Put another way, with a finite number of *types* of claims about observables that have to be sufficiently correct in order to render empirical success, the likelihood of getting those sufficiently right is far lower than the likelihood of getting any other possible results for observables. These considerations strongly suggest that the number of approximately true complexes that will be successful will be lower than the number of approximately true yet unsuccessful complexes. In short, despite the fact that all the statements in our set of complexes are approximately true, empirical failure will be more likely than success. I should point out that, our above considerations pertain to the capacity for an approximately true theory to bring about *general predictive* success. *Novel* success — at least insofar as it is sought in conjunction with general success — constitutes an even greater demand. It is even *less* likely to be brought about by the approximate truth of the theory.

The crucial explanatory premise of realism states that, if a theory is approximately true, then it will (be likely to) enjoy novel success. This premise appears to be false, rendering dubious the claim that the approximate truth of a theory can explain its success. It looks as though the sophisticated realist argument is unable to get off the ground.[60]

Approximate Truth, Reference, and the Meta-Modus Tollens

Let us consider whether sophisticated realism is also subject to the meta-*modus tollens* and its implications. We must ask whether my list above includes theories that were successful but not approximately true. On what grounds might we say that a theory is not approximately true? Laudan draws on the intuition that in order for a statement to be approximately true it must genuinely refer to things in the world. 'If there were no entities similar to atoms, no atomic theory could be approximately true; if there were no subatomic particles, no quantum theory of chemistry could be approximately true' (1981, p. 33). If reference is required for approximate truth, theories that were successful but do not refer cannot be approximately true. Thus theories that have achieved novel success and do not refer (by present lights) will constitute counterinstances to sophisticated realism.

While I think there are a number of reasons for accepting Laudan's claim that the terms of a statement must refer for that statement to be approximately true, I'd suggest the following is crucial. I've just argued that the approximate truth of a theory does not imply or even make it likely that our theory will be successful. This is so even when our notion of approximate truth is one that requires reference. But, were our notion of approximate truth to be applicable to statements that contain altogether nonreferring, *or even partially referring*, terms, it will be even less likely that an approximately true theory will be successful. Eliminating reference from approximate truth threatens to render approximate truth completely non-explanatory. In fact, even appealing to a significantly permissive, charitable notion of reference, e.g., a causal theory of reference, leads to precisely this same problem. Given the list of counterinstances, the realist may be tempted to invoke a notion of approximate truth that allows permissive/charitable reference or no reference at all; however, such a notion of approximate truth will not be compatible with the realist's goal of *explaining* success. For the realist of concern, approximate truth requires reference.

We can now address our list of novel successes as a threat to the sophisticated realist's hypothesis. Construing approximate truth in such a way that we can attribute it to theories containing such terms as 'caloric', 'phlogiston', 'ether', 'absolute space', 'atomic vortices', etc. is not an option for the realist of concern. So the theories including such terms constitute genuine counterinstances even when we make the move to approximate truth. (And since reference is necessary but not sufficient for approximate truth, those theories that can still be construed as referring — in the non-permissive sense required — may still also fail to be approximately true.) Though our list of counterinstances may be decreased with the move to approximate truth, it remains significant. And we are led to the implications of the *meta-modus tollens* discussed in Section Three.

6. AN ALTERNATIVE EXPLANATION FOR THE SUCCESS OF SCIENTIFIC THEORIES?

Let us step back a bit. Our concern here has been with the argument that realism, in some form or another, is the *only* explanation for success. We've now seen that, unless each counterinstance we've considered thus far is inexplicable, another explanation must be available for that theory's success. Here's a contender:

> MS: The mechanisms postulated by the theory would, if actual, bring about all relevant phenomena observed, and some yet to be observed, at time *t*; and these phenomena are brought about by actual mechanisms in the world.

This is my modification of a non-realist contender suggested by Arthur Fine, but which goes back at least as far as Vaihinger. Fine's version — the world is *as if* the theory were true — is deemed *strong surrealism* by Leplin (1987; 1997, p. 26). I call the above version *modest surrealism* (MS). Notice that, provided the phenomena observed are sufficiently wide-ranging, a theory with the property of MS will achieve both general and novel predictive success. The claim that a theory has the property of MS can therefore explain both sorts of success. Moreover, MS expresses a relationship between (a) unobservables in the world and (b) the entities posited by the theory. It also makes causal claims about (a) and (b). Thus, MS is far from being a mere reiteration of the explanandum. On the negative side, MS is predictively vague in that it entails no specification of when the success of the theory will break down. However, the same holds for the claim that our theory is approximately true. Finally, and quite crucially, because MS is not falsified by any of the counterinstances to realism, MS appears to explain the success of every scientific theory we've discussed, while realism gives no explanation at all.

The realist will quickly suggest that, despite the fact that MS goes deeper than reiterating the explanandum of success, we are still left wondering why a theory is modestly surreal. The non-realist replies by pointing out that MS describes something about the unobservable structures of the world and something about the postulates regarding unobservables in the theory: MS specifies that the former does, and that the latter would, bring about the specified phenomena, P. The non-realist will ask, why need we say anything more? The realist will respond, we want to know why *both* have this property. She will then suggest that she has the answer: the theory is approximately true.

In reply, the non-realist will simply point to our considerations above: crucially, we cannot invoke approximate truth to explain why any counterinstances — successful theories that cannot be approximately true — have the property of MS. Second, the approximate truth of T does not even appear to make empirical *success* at *t* likely. And since P goes beyond the phenomena observed at *t*, the approximate truth of T does not make it likely that P will occur. Thus, as with success, the approximate truth of the theory cannot explain MS. On these grounds, the non-realist will contend that if modest surrealism must or even can be explained, it cannot be by appeal to the approximate truth of our theory.

7. DEPLOYMENT REALISM

Given the serious problem with the invocation of approximate truth, the realist may wish to concede that the listed theories are neither fully true, nor approximately true, but suggest that *those constituents of the theories that contributed to success* are true. In other words, the false parts were not responsible for the novel success of the theories. The claim that ether exists, for instance, is not obviously deployed in the derivation of Fresnel's white spot (see Kitcher (1988, pp. 144-149)). A deployment realist hypothesis can be expressed as, 'those constituents that were *used* in the derivation of novel predictions are true'. And the argument meant to justify that hypothesis: it would be a miracle were those constituents not true. In contrast with the approximate truth of the theory, the truth of the deployed constituents *would* make both novel success and MS likely. Thus, this form of deployment realism appears to be a genuine contender for explaining these properties.

In response, one must grant that Fresnel's ether theory is a case in which certain false theoretical claims were not actually employed. However, it is not obvious that many other theories can be broken up in this way. Often the parts of a theory hang together very tightly. I find it highly doubtful that *no* false constituents played a role in the derivation of the predictions on the list.[61] For instance, the numerous novel predictions Mendeleev made are direct consequences of assembling his periodic table. Since that table is a correlation of chemical properties given their atomic weights, it appears undeniable that Mendeleev's periodic law — the claim that atomic *weights* determine chemical properties — played significantly in the construction of that table. To say that absolute acceleration was not used in derivations from Newtonian theory seems implausible. In fact, given the apparent need for absolute acceleration, it is not even clear that Euclidian geometry, absolute space, and absolute time were not involved. Regarding the prediction of the positron, Stephen Brush writes,

> the existence of an anti-particle [ie., the positron] followed directly from [Dirac's relativistic wave] equation, provided one accepted his interpretation of 'holes' in a sea of negative-energy states as particles with the opposite charge and positive energy. But Dirac's theory was replaced by quantum electrodynamics . . . (Brush 1995, p. 139)

Given Brush's account of the history of Benzene theories (1999a, 1999b), one would be hard pressed to deny that Kekulé's theory, now seen as false in light of molecular orbital theory, was employed in the the predictions noted above. Martin Carrier shows that the claim that phlogiston is the principle of heat and the claim that 'sulfuric acid was in fact dephlogisticated sulfur' were directly involved in Stahl's prediction that the synthesis of phlogiston and sulfuric acid would result in sulfur (1993, p. 402). The postulate that charcoal is '[h]igh in phlogiston' and that inflammable air is pure phlogiston, were used in deriving Priestley's prediction that inflammable air would, like charcoal, turn calx into metal. (1991, p. 30.) Following the reasoning of Laplace and Haüy, Carrier also shows that the following constituents, which are now seen as false, were used in predicting that 'the rate of expansion is the same for all gases' (1991, p. 31).

- heat is a weightless fluid called caloric
- the greater the amount of caloric in a body, the greater is its temperature
- gases have a high degree of caloric
- caloric, being a material itself, is composed of particles
- caloric particles have repulsive properties which, when added to a substance, separate the particles of that substance
- the elasticity of gases is caused by this repulsive property of caloric heat particles
- the elasticity of all gases is due to the repulsive properties of a single substance, caloric

Psillos acknowledges the following regarding Poisson's prediction of the adiabatic law: 'To be sure, Poisson did *rest* his derivation' of the law of adiabatic change on the state function hypothesis (1999, p. 120), [my italics] which is the 'fundamental hypothesis of the mature caloric theory' (1999, pp. 120-121). However, the state function hypothesis requires that heat is always conserved, and, by present lights, heat is not always conserved.[62]

Keith Hutchison has given a detailed history of W. J. M. Rankine's (often overlooked) mid-nineteenth century vortex theory, detailing its exceptional success.[63] This theory made at least four temporally novel predictions plus half a dozen or so use-novel predictions.[64] The beauty of Rankine's theory, in regard to deployment realism, is that Rankine himself sought diligently to achieve the same results while eliminating as much of his vortex theory as possible. Despite his efforts, however, he had to make explicit appeal to the vortex hypothesis. As Hutchison puts it, 'Rankine's "phenomenological" theory . . . remained logically dependent on the vortex theory' (1981a, p. 13). And Rankine's theory is dramatically false by present lights. This theory appears to provide exceptional counterevidence to the realist's suggestion that the false parts of theories were not used in the derivations of novel predictions.

Dalton derived his law of partial pressure from an early version of his (false) atomic theory in 1801. More specifically, as John Hudson points out, he 'deduced' it from 'the idea that each particle could only repel others of its own kind, and that dissimilar particles exerted no forces on each other' (1992, p. 80). Dalton's law of multiple proportions was also, as James Partington puts it, 'a consequence of his Atomic Theory' (1957, p. 159). That law states

> when two elements combine to form more than one compound, the weights of one element which unite with identical weights of the other are in simple multiple proportions. (Partington 1957, p. 159.)

H.M. Leicester notes that this law 'followed so essentially from Dalton's theory that he did not even express it as a distinct principle.' (1965, p. 155). Among those specific claims that were crucially employed in the derivation of the law of multiple proportions, and thereby more specific novel successes, is Dalton's principle of simplicity:

> where two elements A and B form only one compound, its compound atom contains one atom of A and one of B. If a second compound exists, its atoms will contain two of A and one of B, and a third will be composed of one of A and two of B, etc.
> (Hudson 1992, p. 81)

This principle, however, is false by present lights.[65] Yet it was — at times by way of its deployment toward the law of multiple proportions — genuinely employed in numerous confirmed temporally novel predictions (see the list in Section Four). Going beyond our list of novel successes, it should be recognised that thousands of previously unknown chemical compounds have been predicted by way of chemical principles such as Proust's law of definite proportions and Dalton's law of multiple proportions. Yet, as Maureen Christie points out (in a different context), these principles cannot be regarded as strictly speaking true. Each is 'exceptioned' and/or approximate, at best (1994, pp. 613-619).

I must also reemphasize that the list above includes only examples of *temporally* novel success. Yet the realist (in following Zahar's discussion from a separate debate) commonly appeals to use-novelty as well. When a theory complex makes a prediction after a new hypothesis is added, and was not *seen* to make that prediction before, it is clear that the new hypothesis is genuinely employed in predicting the phenomena. Noting this, we see that accommodating hypotheses with little external support can very well be and often have been use-novel. Within the context of phlogiston theory, the claim that phlogiston has negative weight was introduced to explain why calxes are heavier than their metals. Once formulated, however, that hypothesis could also be deployed in the use-novel prediction that fire behaves differently than other substances: because fire has negative weight, it moves away from massive bodies rather than toward them. In order to reconcile the discrepancy between the behavior of Mercury and Newtonian mechanics, it was proposed that a collection of invisible planetoids, small planets, serve to offset Mercury's perihelion. However, that postulate also predicted, in the use-novel sense, future instances of the glow in the sky observed post-sunset on the western horizon and pre-sunrise on the eastern horizon, known as 'the zodiacal light'. Astronomer Simon Newcomb wrote: 'if the group exists the members must be so small as to be [individually] entirely invisible. But in this case they must be so numerous that they should be visible [collectively] as a diffused illumination on the sky after sunset. Such an illumination is shown by the zodiacal light'.[66]

Now in contrast with most discussions about use-novelty, our concern is truth and justified belief, rather than rational theory choice. That given, coupled with the fact that use-novelty entails no temporal restrictions, ancient theories that make use-novel predictions of modern phenomena would have to qualify as well.[67] Even setting aside such anachronistic cases, since use-novelty is very common in science, we can likely find many more historical examples of false constituents being deployed toward use-novel predictions. For instance, Newtonian mechanics predicted the behaviour of the tides, the behaviour of unseen stellar objects, the precession of the equinoxes; Bohr's theory of the atom predicted the Balmer series of the hydrogen spectrum, etc. Rankine's vortex theory achieved a significant amount of *use*-novel success as well. I submit that, upon a close search, the list of counterinstances to deployment realism would be very substantial.

This final stand in the series of retreats I've traced is, in a very significant way, a bolder position than the others: there are far more potential falsifiers for this thesis than for the more broad-based theory realisms. Many constituents will be involved in the theories listed above, and each false *constituent* that is employed in a confirmed novel prediction constitutes a counterinstance to deployment realism. Thus many counterinstances can come from a single prediction. For example, it looks as though we have at least seven false constituents employed in Haüy and Laplace's reasoning toward the prediction that 'the rate of expansion is the same for all gases' (Carrier 1991, p. 31). Further, the deployment realist's hypothesis pertains to auxiliary assumptions no less than broad ranging theoretical claims. It seems altogether untenable to claim that every constituent, invoked explicitly or implicitly, that was deployed in the predictions we've discussed is true. Moreover, a given false constituent stands as a separate counterinstance *each time* it is employed in the derivation of such a prediction. Deployment realism has the potential to fare far worse than most versions of realism given the historical argument. Consider Dalton's principle of simplicity. Or better yet, Mendeleev's periodic law: this law was genuinely employed in, at least, those dozen predictions listed in Section Four — and that number is setting aside his prediction of individual *properties* of each of the predicted elements (see my footnote on Mendeleev in Section Four). Thus, that law would stand as a counterinstance at least a dozen times, once for each of the predictions drawn from it. Consider as well the use of the laws of definite and multiple proportions in the predictions of thousands of chemical compounds. My bet is that on close examination, the falsifying instances for this final stand of theory realism will be far more numerous than they are for the more naive realisms considered above. Though further in depth analysis is needed, suffice it to say that each of the claims above appears to be genuinely employed in novel predictions, yet each is strictly speaking, if not dramatically, false by present lights. Wc are again led to the implications of the *meta-modus tollens* in Section Three.

Regarding the ability to explain modest surrealism (MS), deployment realism affords no explanation for novel successes from false constituents. Nor, then, can it explain why the respective broader theories have the property of being modestly surreal. Though MS may have less *depth* than the realist desires, it has an incomparable degree of *explanatory breadth*: MS explains the success of both the general predictive success and the novel success of, not only every theory mentioned in this paper, but all that we've neglected, including the corpus of today's science. In this sense, it is a far superior explanation for the success of scientific theories than any version of realism we've here considered.[68]

8. CONCLUSION

The following number among our conclusions. The threatening historical argument is neither an induction nor an inference to the falsity of present theories, and numerous successful novel predictions have come from false theories. It looks as though the approximate truth of T does not make T likely to succeed. And a nonreferential notion of approximate truth is in conflict with the explanationist argument for realism. We've seen further that, given the counterinstances at each level, there are a number of significant implications of the meta-*modus tollens*. False theories are rendered miracles; thus the first premise of the realist's inference is unacceptable. And, despite the realist's own insistence on the need for explanation, the much-advertised explanatory ability of realism is rendered dubious given its inability to explain these instances. The realist hypotheses — the asserted correlations between success and truth, approximate truth, etc. — have no confirming instances without presupposing realism. And each realist hypothesis is falsified many times in the history of science. Finally, we've seen a non-realist competitor that, in great contrast to realism, explains the success of all successful scientific theories. Though I consider myself to be a scientific realist of sorts, a theory-based scientific realism invoking (approximate) truth has yet to make its case.

Indiana University—Purdue University Indianapolis (IUPUI)

REFERENCES

Brown, J.R.: 1985, 'Explaining the Success of Science', *Ratio* 27, 49-66.

Brush, S. G.: 1995, 'Dynamics of Theory Change: The Role of Predictions', *PSA 1994*, vol 2. Hull, Forbes, and Burian (eds), Philosophy of Science Association, East Lansing, MI, pp. 133-145.

Brush, S. G.: 1999a, 'Dynamics of Theory Change in Chemistry: Part 1. The Benzene Problem, 1865-1945', *Studies in History and Philosophy of Science*, vol. 30, no.1, 21-79.

Brush, S. G.: 1999b, 'Dynamics of Theory Change in Chemistry: Part 2. Benzene and Molecular Orbitals, 1945-1980', *Studies in History and Philosophy of Science*, vol. 30, no.2, 263-302.

Carrier, M: 1991, 'What is Wrong with the Miracle Argument?' *Studies in History and Philosophy of Science*. vol 22, 23-36.

Carrier, M.: 1993, 'What is Right with the Miracle Argument: Establishing a Taxonomy of Natural Kinds' *Studies in History and Philosophy of Science*, vol 24, no.3, 391-409.

Cartwright, N: 1983, *How the Laws of Physics Lie*, Oxford: Oxford University Press.

Christie, M: 1994, 'Philosophers versus Chemists Concerning "Laws of Nature"', *Studies in History and Philosophy of Science*, vol 25, no 4, 613-624.

Clarke, S.: 2001, 'Defensible Territory for Entity Realism', *British Journal for the Philosophy of Science* 52, 701-722.
De Regt, H: 1994, *Representing the World by Scientific Theories: The Case for Scientific Realism*. Tilburg University Press, Tilburg.
Devitt, M.: Forthcoming, 'Scientific Realism' (draft), F. Jackson and M. Smith, (eds.), *The Oxford Handbook of Contemporary Analytic Philosophy*.
Duhem, P.: 1906, *The Aim and Structure of Physical Theory*. P. Wiener (trans), Princeton University Press, Princeton (1954).
Ellis, B.: 1979, *Rational Belief Systems*, Blackwell, Oxford.
Ellis, B.: 1990, *Truth and Objectivity*, Blackwell, Cambridge.
Greenstein, G.: 1988, *The Symbiotic Universe: Life and the Cosmos in Unity*, Morrow, New York.
Guth, A.: 1997, *The Inflationary Universe*. Reading, Addison-Wesley, Massachusetts.
Hacking, I.: 1983, *Representing and Intervening*. Cambridge University Press, Cambridge.
Hudson, J.: 1992, *The History of Chemistry*, Macmillan Press, London.
Hutchison, K.: 1973, *The Rise of Thermodynamics*, paper read at the 1973 *Australasian Association for the History and Philosophy of Science Conference*, the University of Melbourne.
Hutchison, K.: 1981a, 'W.J.M. Rankine and the Rise of Thermodynamics', *The British Journal for the Philosophy of Science*, 14, no. 46.
Hutchison, K.: 1981b, 'Rankine, Atomic Vortices, and the Entropy Function', *Archives Internationales d'Histoire des Sciences*, 31, 72-134.
Isaacs, A, et al.: 1991, *Concise Science Dictionary*, 2nd edition, Oxford University Press, Oxford.
Kitcher, P.: 1993, *The Advancement of Science*, Oxford University Press, Oxford.
Knight, D.: 1989, *A Companion to the Physical Sciences*, Routledge, New York.
Lakatos, I.: 1970, 'Methodology of Scientific Research Programmes', in I. Lakatos and A. Musgrave (eds), *Criticism and The Growth of Knowledge*, Cambridge University Press, Cambridge.
Laudan, L.: 1977, *Progress and its Problems*, Oxford University Press, Oxford.
Laudan, L.: 1981, 'A Confutation of Convergent Realism', *Philosophy of Science*, 48, 19-49.
Leicester, H.M.: 1965, *The Historical Background of Chemistry*, Wiley, New York.
Leplin, J.: 1987, 'Surrealism', *Mind*, 96, 519-524.
Leplin, J.: 1997, *A Novel Defense of Scientific Realism*, Oxford University Press, Oxford.
Lipton, P.: 2000, preprint, 'Tracking Track Records', for *Proceedings of the Aristotelian Society Supp*. Vol. LXXIV.
Marks, J.: 1983, *Science and the Making of the Modern World*, Heinemenann Press, London.
Mason, S.: 1962, *A History of The Sciences*, Macmillan, New York.
Motz and Weaver.: 1989, *The Story of Physics*, Avon, New York.
Musgrave, A.: 1985, 'Realism versus Constructive Empiricism', in P. Churchland and C. Hooker (eds), *Images of Science*, Chicago University Press, Chicago.
Partington, J.: 1957, *A Short History of Chemistry*, Macmillan, London.
Peirce, C. S.: 1958, *Collected Papers*, vol. 5, Harvard University Press, Cambridge.
Poincaré, H.: 1902, *Science and Hypothesis*. Dover, New York (1952).
Psillos, S.: 1999, *Scientific Realism: How Science Tracks Truth*, Routledge Press, London.
Putnam, H.: 1984, 'What is Realism', in Leplin (ed), *Scientific Realism*, California University Press, Berkeley.
Putnam, H.: *Meaning and the Moral Sciences*, Routledge Press, London.
Rescher, N.: 1987, *Scientific Realism: A Critical Reappraisal*, D. Reidel, Dordrecht.
Sankey, H.: 2001, 'Scientific Realism: An Elaboration And A Defense', *Theoria*, 98, 35-54.

Scerri, E.R.: 2000, 'Realism, Reduction, and the "Intermediate Position"', in N. Bhushan. and S. Rosenfeld (eds)., *Of Minds and Molecules*, Oxford University Press, New York.
Van Fraassen, B. C.: 1980, *The Scientific Image*, Oxford University Press, Oxford.
Worrall, J.: 1984, 'An Unreal Image', *The British Journal for the Philosophy of Science*, 35, 81-100.
Worrall, J.: 1989a, 'Structural Realism: The Best of Both Worlds?' *Dialectica*, 43, 99-124. Reprinted in, 1996, *Philosophy of Science*. D. Papineau (ed), Oxford University Press, Oxford.
Worrall, J.: 1989b, 'Fresnel, Poisson and the White Spot: The Role of Successful Predictions in the Acceptance of Scientific Theories', *The Uses of Experiment: Studies of Experimentation in Natural Science*. D. Gooding, T. Pinch and S. Schaffer (eds.), Cambridge University Press, Cambridge.
Worrall, J.: 1994, 'How to Remain (Reasonably) Optimistic: Scientific Realism and the "Luminiferous Ether"', *PSA 1994*, vol 1. M.Forbes and D.Hull (eds), Philosophy of Science Association, East Lansing, MI.

NOTES

[1]In this paper, I am restricting my discussion to truth-based theory realism and am setting aside the entity realisms of Ellis (1979) (1990), Hacking (1983) Cartwright (1983) and Clarke (2001).
[2]Since the attribution of miracles is seen to be equivalent to giving no explanation at all, we can read this argument as claiming that the theory's truth is the *only possible* explanation for predictive success.
[3]While knowing that our theories are likely to be false would bear on whether or not we are *justified in believing* that they are true, this is not Laudan's approach.
[4]Genuine versions of the pessimistic meta-induction can be found in Rescher (1987, p. 5) and Putnam (1984, p. 147). In fact, Laudan himself did state a version of the pessimistic induction four years earlier (1977, p. 126). Nonetheless, I submit that Laudan is making no pessimistic meta-induction in the 'Confutation' article.
[5]While Psillos refers to Laudan's list as an inductive inference, he also puts an appropriate spin on it (1999, p. 99, p. 102). However, he is inconsistent in this, to the detriment of his own defence against Laudan. His wavering makes salient the fact that these two arguments are often conflated.
[6]I've recently read a preprint by Peter Lipton in which he presents a careful analysis that details the precise factors that render this induction weak (2000).
[7]While, on occasion, Laudan's argument appears to be *implicitly treated* as I've construed it, in order to appreciate the full implications of the argument (to be spelled out below), I consider it important that the proper formulation be made explicit.
[8]Jarret Leplin makes this point against Laudan (again, misconstruing Laudan's argument as a pessimistic induction) (1997, pp. 141-142). See also Devitt (forthcoming, p. 19, footnote, 30), who appears to share Leplin's concern.
[9]Laudan does speak of the theories on his list as being 'evidently false' (1981, p. 35) and false 'so far as we can judge' (1981, p. 33). In order to keep the argument clean, he should instead only say they are false 'according to present science', 'by contemporary lights', etc.
[10]Putnam (1984, pp. 143-144) makes this move, albeit, in his more sophisticated realism.

[11]In the same breath, realists may introduce the claim that it is possible to justifiably believe something that we later conclude to be false. This, however, would serve as no response to the meta-*modus tollens*. The counterinstances challenge the justification for believing the realist hypothesis *in the first place*, irrespective of whether a particular theory is later concluded to be false.

[12]Novel success is also appealed to in response to van Fraassen's Darwinian explanation of the success of science. See Worrall (1984), Brown, (1985), and Musgrave (1985, 1988).

[13]Carrier (1991, pp. 30-31)

[14]Hacking (1983, p. 69)

[15]Hutchison (1973)

[16]Knight (1989, p. 120); Hutchison (1973)

[17]Psillos (1999, p.120). This is an appropriate place to acknowledge, along with Whewell, Duhem, Psillos, *et al*, the great significance of the derivation of novel *theoretical generalisations* as well as *singular empirical* predictions. In fact, I think few would disagree that, so long as such predicted theoretical generalisations are retained in a later theoretical scaffolding, they are even *more* valuable than singular empirical predictions. For from the predicted theoretical generalisations, one can derive further individual empirical predictions. And, so long as the empirical successes of those generalisations don't rest on assumptions that contradict the original theory, these empirical predictions count no less as successes for the original theory.

[18]Carrier (1993, p. 402)

[19]Carrier (1991, p. 30)

[20]Hutchison (1981a, p. 9)

[21]Hutchison (1981a, p. 9)

[22]Hutchison (1981a, p. 8). Rankine named this state-function 'the thermodynamic function' in 1854. But it was already present in his papers read on February 2, 1850. As Hutchison explains, 'All the involved calculation had been done by 1850, though Rankine still lacked a motive to single out the entropy function for specific attention' (p. 8). This was 15 years before Clausius's claim that entropy increases.

[23]Hutchison (1981a, p. 9). See also Hutchison (1981b, p. 104)

[24]Mason (1962, pp. 289-290)

[25]Mason (1962, pp. 292-293)

[26]Worrall (1989a, p. 142)

[27]Motz and Weaver (1989, p. 104, p. 123)

[28]Worrall (1989a, 1989b, 1994)

[29]Worrall (1989b, p. 145)

[30]Worrall (1989b, p. 152). See also the comments of Arago and Poinsot quoted in Worrall (1989b, p. 143)

[31]Worrall (1994, p. 335)

[32]Mason (1962, pp. 485-6)

[33]I've blended at least four confirmed predictions together here. Thomson found specifically, that acid salt contains exactly half the oxalate of normal salt. He found the same result when the base being held constant is strontium (Hudson, 1992, p. 84); (Leicester, 1965, p. 156); (Partington, 1957, p. 159). Partington states that Thomson varied the base rather than the

acid, but this is not important here. Wollaston added a third compound to Thomson's results, and found the three to have a 1:2:4 ratio. This third salt being potassium tetroxaliate (Hudson, 1992, p. 84); (Partington, 1957, p. 159). Just as with the oxalates, Wollaston found a 1:2 ratio between the quantities of the carbonic acid in the different carbonates. Likewise, the acid in the two sulfates was again in a 1:2 ratio (Partington 1957, p. 159).

[34]These confirmed predictoins came after his theory was put forward in 1803 (see Partington 1957, p. 158, p. 172); (Hudson 1992, p. 82).

[35]He arrived at this law from an early version of his atomic theory in 1801. He 'deduced' it from 'the idea that each particle could only repel others of its own kind, and that dissimilar particles exerted no forces on each other' (Hudson 1992, p. 80).

[36]Regarding the last four see Hudson (1992, p. 82).

[37]Hudson (1992, p.83)

[38]Specifically, in 1874, Landenburg showed there is no second isomer in pentachlorobenzene (Brush, 1999a, p. 25).

[39]Brush (1999a, pp. 27-28); (1999b, p.264)

[40]Brush (1999a, p. 32)

[41]Brush (1999a, pp. 58-61); (1999b, pp. 265-6, p. 290)

[42]This was confirmed in 1874 (see Leicester, 1965, p. 95). In fact, when Gallium was first discovered, its specific gravity was found to be a bit lower than Mendeleev had predicted. Mendeleev asked de Boisbaudran to recheck the value, and Mendeleev's value was confirmed (see Marks 1983, p. 313).

[43]This shorthand of merely *naming* these elements constitutes a great injustice to Mendeleev's actual predictions. In predicting germanium (which he called eka-silicon), for instance, what Mendeleev did was predict the existence of an element with the following properties: an atomic weight of 72, confirmed at 72.32; a specific gravity of 5.5, confirmed at 5.47; an atomic volume 13, confirmed at 13.22; a valence of 4, confirmed at 4; a specific heat of 0.073, confirmed at 0.076; a specific gravity of dioxide of 4.7, confirmed at 4.703, a molecular volume of dioxide of 22, confirmed at 22.16, plus many other such properties. (See Tables 1 and 2 in Leicester 1965, pp. 195-196). Mendeleev made amazingly detailed predictions for each of the first three elements listed here.

[44]Confirmed in 1879 (Leicester 1965, p. 195).

[45]Confirmed in 1885 (Leicester 1965, p. 196).

[46]That Mendeleev predicted these last four was pointed out to me by John and Maureen Christie. In the literature on Mendeleev, these are less commonly discussed than the first three. To see that he predicted these one can compare Mendeleev's table with the contemporary table. See, for instance, tables in Hudson (1992 p. 133, p.137). On Perrier and Segre's confirmation, see Isaacs (1991, p. 682).

[47]Confirmed in 1925 (Greenwood and Earnshaw 1986, p. 1211).

[48]Confirmed in 1939 (Isaacs 1991, p. 274).

[49]Confirmed in 1898 (Isaacs 1991, p. 544).

[50]Scerri (2000, p. 59)

[51]Scerri (2000, p. 59, footnote 27)

[52]Regarding the preceding five predictions (collapsed to three): Mendeleev's table reorders these elements, correctly contradicting the data then available; his corrections were later

confirmed. (See Hudson and compare the tables on 128 (data available) with 129 (Mendeleev's table)).

[53]Lakatos (1970, p. 147)

[54]Mason (1962, p. 558, p. 560)

[55]Guth (1997, Ch. 4)

[56]Peirce's construal of abductive reasoning requires that an explanation (A) makes a surprising fact (B) 'a matter of course' (1958, p. 189). I'd suggest that premise 1 is a liberal interpretation of this claim.

[57]Realists tend to neglect this issue. For instance, while Stathis Psillos attempts to elucidate just what verisimilitude is (Chapter 11, 1999), at no point there does he argue that such a property will even render empirical success likely.

[58]Notice that in pointing out that this consequence would follow from C2-Cn, no assumptions are being made about the world. It is known analytically given the corpus of science: we need only look at what the theories say to see that they make this prediction. (Of course, pointing out that such a prediction would fail empirically rests on the basic empirical claim that there is a universe.) The consequence of changing the charge of the electron, etc., is discussed in, for instance, Greenstein (1988, pp. 61-65).

[59]It should be clear that the result is not symptomatic of the fact that I've used the corpus of science. We can delimit our theory surrounding these changes in any number of ways and achieve the same result. Pre-reflectively, one might be concerned that my example fails for the following reason: the corpus of science simply prohibits the possibility that any such slight change in these charges can occur, thus, the alternatives cannot really be approximately true. (The claim internal to C1 that generates this concern would be the claim that a slight modification in charges causes matter to repel.) However, we cannot look to a given theory itself in order to determine what would make that theory approximately true. For the purposes of realism, our notion of approximate truth must, of course, be defined independently of any particular theory.

[60]If the realist argument does not involve Peirce's premise, it is most pressing that realists clarify just what their replacement premise is and then show that that premise holds in light of the objections above.

[61]Psillos attempts to show those claims that were 'essential' to 'key predictions' have been retained across theory change. The criterion he presents for determining what he calls the 'essentiality' of a postulate is very elaborate (1999, p. 110). However, he himself does not appear to take that criterion seriously. Nor will I here. I have pointed out elsewhere that he does nothing to show that retained theories fit his criterion or that rejected theories fail to fit his criterion. In fact, after introducing his criterion, in the 36 some odd pages where he is addressing the historical argument, he never employs or even mentions that criterion again. Moreover, his criterion is not suitable to the purposes for which he has invoked it — namely, determining which constituents are *responsible* for the prediction; it is too vague to be applicable; and it does not appear to be epistemically motivated. In fact, he inadvertently replaces it with what turns out to be an altogether different criterion for realism: we should be realists about the constituents to which scientists are (or at least express themselves in print as being) committed.

[62]Though a deployment realist, Psillos can admit this for his purposes, as he is attempting to show that the constituents were not *essential*, which he defines using a much stricter condition than the mere *use* of a constituent. See footnote above.

[63]As detailed by Hutchison in (1981a).

[64]Notably the richness of Rankine's theory is not exhausted by its temporally novel and use-novel success. For instance, Rankine's was a unifying theory, exhibiting a significant degree of breadth, reconciling seemingly disparate phenomena. Hutchison writes, 'unlike earlier proponents of this general view, Rankine extended the hypothesis to cover the structure of the luminiferous aether as well as that of ordinary matter' (1981a, p. 4).
[65]It leads for instance to Dalton's conclusion that water is HO rather than H_2O.
[66]I have borrowed Newcomb's quote from Adolf Grünbaum ([1976], quoted on p. 356). Grünbaum is concerned with a different issue, however, namely, *ad hoc* hypotheses in science and their implication for Popper's system.
[67]For instance, Aristotelian cosmology may well be seen to predict, in a use-novel sense, the rise of certain modern objects that appear to be fire driven — e.g., hot air balloons, bottle rockets, the space shuttle, etc.: the fire seeks to return to its natural place, and as its quantity is greater than the quantity of earthly matter in the given object, it pushes that object upward.
[68]I am interested in rendering MS more robust. I consider promising the possibility of conjoining it to variants of the taxonomical realism of Duhem (1906) and Carrier (1991, 1993) and/or the structural realism of Poincaré (1902) and Worrall (1989a).

KEITH HUTCHISON

MIRACLE OR MYSTERY? HYPOTHESES AND PREDICTIONS IN RANKINE'S THERMODYNAMICS

1. OUR PROBLEM AND ITS CONTEXT

This paper sketches an historical episode that presents a major obstacle to the project of using sustained empirical success as a warrant of truth. Its target is the 'no miracle argument', used by a family of realist philosophies of science to infer that significant success in deriving some features of the world from a theory provides a very strong reason to believe in that theory. For if the theory in question is true, or approximately so, such success is no mystery. But if the theory is seriously false, success is some sort of coincidence, especially so in the case of novel predictions. Some see the coincidence as huge, almost a miracle, and conclude that when a significant prediction succeeds it is much better to believe the theory behind it is true.[1]

I argue the contrary case, that such coincidences are not so rare. I do this incompletely, via a single family of examples, by examining Rankine's mid-19c kinetic model of heat, a model that no-one today is likely to deem even approximately true. I scrutinise Rankine's ability to explain a whole range of important thermal phenomena, and its uncanny ability to generate successful predictions — a whole series of them, of great importance to the early development of thermodynamics (including, indeed, the existence of entropy). Such successes were of course evidence in favour of the theory, but in the end, the evidence proved misleading.

So I use Rankine's thermodynamics in much the same way as others have used the Ptolemaic theory, Newton's mechanics, or the caloric theory, etc. in what has been called a 'pessimistic induction', to indicate that the coincidences at issue do indeed occur, quite frequently.[2] But my episode seems more robust than those others. For it is quite plausible to suggest that Ptolemaic astronomy worked precisely because it contained a core of timeless truth, combined with a plethora of more or less irrelevant falsehoods: the orbits Ptolemaic theory allocates to planets can be reinterpr-

S. Clarke and T.D. Lyons (eds.), Recent Themes in the Philosophy of Science, 91–120.

eted as algorithms for the heliocentric calculation of *relative* motion, and as *approximations* furthermore to Kepler ellipses.[3]

To show that Rankine's theory is quite different is the main aim of my discussion. The heart of the analysis will be a close scrutiny of the reasoning that led Rankine to make his predictions. I will show that his reasoning depends vitally on unusual features of the microscopic motion he held responsible for thermal phenomena. These features are so incompatible with modern views, that there is no temptation to allocate any sort of approximate truth to them. If we know *any* theory is false, we know that Rankine's is! So Rankine's work provides a compelling example of false theory that is remarkably successful, especially at novel predictions. It refutes the suggestion that a miracle is needed for this to happen.

Oddly, the conclusion we reach will turn out to be somewhat stronger. For we will eventually observe that Rankine's successes are not just the consequence of a false *theory*, but seem to rely on defective *reasoning* as well — rather like Maxwell's prediction that light is electromagnetic (Siegel 1991, pp. 137-8), or Le Verrier's discovery of Neptune (Pannekoek 1961, pp. 360-1). So it is even less plausible to seek to defuse the conclusion, by some variant of the 'Ptolemy-argument' sketched above. Of course, the episode then becomes a problem for *all* empirical philosophies of science. *However* a philosopher might articulate the virtues of a 'good' theory, those virtues cannot be attached to a theory in consequence of the confirmation of novel predictions — until we check that the theory did in fact justify the prediction. But this problem is relatively mild for some: a Bayesian can, for instance, simply introduce a probability for the possibility that the reasoning behind some prediction is invalid, and his calculations might well show that the probability a theory is only slightly increased by a successful prediction when it is discovered that this probability is very high. But the conclusion drastically undermines the no miracle argument. For it simply makes the hypotheses that success has occurred by coincidence even more acceptable.

2. RANKINE'S THERMAL HYPOTHESIS

The early history of thermodynamics hinges on two critical dates: 1824 and 1850/1.[4] The first of these years saw the appearance of Carnot's amazingly penetrating analysis of the efficiency of heat-engines.[5] and the second saw Clausius's (then Thomson's) resolution of the intellectual puzzle created by the clash of Carnot's theory with the newly emerging idea that heat and work were inter-convertible. Among the successful predictions made by Clausius's theory is one that is not just novel but positively paradoxical: saturated water vapour has a *negative* specific heat![6] For complicated reasons, Thomson saw this result as confirmed by the fact that escaping steam tends not to scald.[7] But oddly, Thomson actually met the result somewhere else, before he found it in Clausius. For the same prediction had been announced by the Scottish engineer, W.J.M. Rankine, three months before Clausius.

Unlike the thermodynamics of Thomson and Clausius, that developed by Rankine was heavily dependent on a characteristic view of the microstructure of matter.[8] According to this view, matter consisted of a scattering of nuclei (whose nature and operations will matter little below) in a motion-filled 'atmosphere'. A nucleus together with the (approximately spherical) portion of atmosphere that it dominated was referred to by Rankine as an 'atom'. We will use the same terminology, but it is important to realise such things were not atoms in the classical sense. For they were divisible (in that they possessed component parts — nucleus and atmosphere, for instance); and some atmosphere was perhaps shared between neighbouring nuclei.

As an atmosphere-nucleus model of the atom, Rankine's is certainly one of the historical sources of current beliefs, but it is also very different. In particular, its account of heat is radically different, for thermal behaviour results primarily from the motions inside Rankine's atoms — within their atmosphere — and is not in any way related to collisions (either of atoms, or of anything else). Furthermore, there is no use made of any irregularity or disorder in those motions. Indeed the motions are reasonably well-organised and such minor disorder as Rankine happens to allow is shielded from generating any thermal effects.[9] So Rankine's hypothesis uses an arrangement of motion that no-one today imagines to be remotely connected with heat. It follows that any predictions that genuinely depend on this model of matter are derived from a seriously false theory of heat.

Rankine's theory seems to have developed around 1840 in an attack on the then contemporary problem of developing an account of elasticity adequate to explain the motions of light through the aether. Rankine's nuclei indeed act on each other via action-at-a-distance forces, and their motions, coordinated by these forces, were hypothesised to constitute light. The interactions between them however were modified by immersion in the atmosphere mentioned above, a fluid (not the aether!) occupying the gaps. This fluid is a little like a perfect gas for it has an *Ur*-elasticity, a primitive tendency to expand, and the pressure (p) it exerts as it seeks to do this is always proportional to its density (D, with $p = bD$). In other words, the atmosphere obeys a primitive version of the Boyle-Mariotte law (that which asserts PV to be constant in a sample of gas). Rankine does not attempt to explain this innate elasticity: it is what he assumes, to make sense of other phenomena. But it is of different strength in different chemicals.

The key to his thermodynamics is the motion Rankine allocates to this fluid, for such motions are the ones which 'constitute' heat. Somewhat akin to more modern theories, their kinetic energy is taken as the measure of heat;[10] and the forces they generate are the source of the mechanical effects produced by heat. Rankine however does not expound the characteristics of these motions too clearly, in part because he sought a theory that was not tied to particular details, and perhaps because contemporaries already knew what he had in mind.

But Rankine leaves no doubt about what really matters. The fluid in an atom is *not* in motion about the nucleus, but instead, each portion of it rotates about a nearby *radius* centred on the nucleus. The atoms indeed are filled with innumerable tiny tapered cylinders, one of which is depicted in fig.1. So we can perhaps envisage each atom as filled with rotating filaments, threads emanating from the nucleus, and perhaps connecting nuclei of different atoms together, but such an elaboration is not essential. The only thing that matters is the fact that such rotating cylinders change the forces that an atom exerts on its environment. A good analogy is provided by water in a bucket: when this is in rotation, the water on the outside is slightly more compressed than when it is stationary, and the force exerted by the water on the walls of the bucket is accordingly somewhat greater. Such motion-generated forces are the key to Rankine's thermodynamics, for they are the means by which macroscopic objects modify their mechanical activity in response to thermal stimuli, so in the end, they provide the means by which heat is harnessed to produce useful work. And to develop his thermodynamics, Rankine calculates the way they behave, via quite unsophisticated mechanics.

3. A PRELIMINARY ORIENTATION

But the mechanics is somewhat complicated, and the relationship between the various parts of the argument below is rather hard to grasp, so let us begin with a bird's-eye view of the analysis. Rankine ends up with five obvious types of empirical success (all predictions in some reasonable sense, examined more closely below), viz:

(A) *Imperfect Gasses*: in 1850, he develops (remarkably accurate) forms of the equation of state for both steam and imperfect gases (as empirically observed in the late 1840s) — from theoretic results reached in the early 1840s.

(B) *Cooling Effect*: the 1850 equation of state (of ¶A above) is used (in 1854) to calculate the cooling experienced by carbon dioxide in the Joule-Thomson expansion experiments (of about the same time).

(C) *Entropy*: he predicts the existence of the entropy function: explicitly in 1854, but implicitly in 1850.

(D) *Specific Heat (Gases)*: in 1850 he predicts that the specific heats of perfect gases will be constant, and perhaps[11] predicts the actual value of one of them. These facts were confirmed in 1853.

(E) *Specific Heat (Steam)*: in 1850, he makes the prediction already noted — and soon accepted[12] — that saturated steam has a negative specific heat.

To reach these results, Rankine uses his model of thermal motions (that just sketched) to perform three core calculations (all announced in 1850), viz.:

(i) *Pressure Calculation*: yields the pressure at the boundary of a Rankine-atom in terms of the speed of vortical rotation.

(ii) *Temperature Calculation*: Links temperature to the speed of vortical rotation.

(iii) *Interaction Calculation*: Tells what happens when external forces compress the atom slightly.

These core calculations contain all the hard work involved in Rankine's predictions. Once they are reached, he quickly observes their immediate macroscopic consequence — by doing little more than noting he has assumed there are no gaps between the atoms in a material system. Then various forms of juggling (some are a bit complicated in logical structure — fig. 3, will give us an overview) lead to the results listed above. So let us now move on to the details. Our principal task here is to check that the false motions are genuinely implicated in making the predictions above: this task is a little tedious, so I summarise the results reached later in the paper.

4. THE MICROSCOPIC CALCULATIONS

4.1. *The Pressure Calculation.*[13]

In this calculation, Rankine finds a relationship between the pressure (P) exerted by the atom of some gas on its environment; the atom's density (D); and the speed of vortical rotation — which is deemed to have constant linear velocity (w) all through the gas. Treating the atom as a sphere, Rankine finds[14]

$$P = bD.(1+w^2/3b).(1-\xi(D,w)),$$
[equation 1]

where $\xi(D,w)$ is function associated with the gas, representing the effect of nuclear action-at-a-distance on the atmosphere. It becomes small as D becomes small, and as w becomes large according to the form

$$\xi(D,w) = A_1(D)/w^2 + A_2(D)/w^3 + ...$$
[equation 2]

where the A_i are functions that become small as D does and as i increases.

The crux of the calculation amounts to a modelling of the atom by a macroscopic sphere of a special almost-perfect gas. Imagine a sphere full of perfect gas, attracted towards the centre by an action-at-a-distance force, so that its density increases towards the centre. Calculating the way the density changes is not a difficult task for those who have done a little calculus. Let us change the problem slightly however, and suppose the gas is not isotropic, in that pressure exerted *at right angles to* radii is greater than that exerted *along* the radii — as indicated in fig.1, where the tapered cylinder is acted on by (i) radial pressures at each end, (ii) the larger transverse pressure on the curved sides, and (iii) the central force. *Such a sphere now models Rankine's atom*, for the fact that his atom is filled with cylinders rotating along their axes generates the additional transverse force imagined in our model of the atom. Again we can do a calculation, and we find that the pressure at the sphere's boundary is that given by eq. 1.

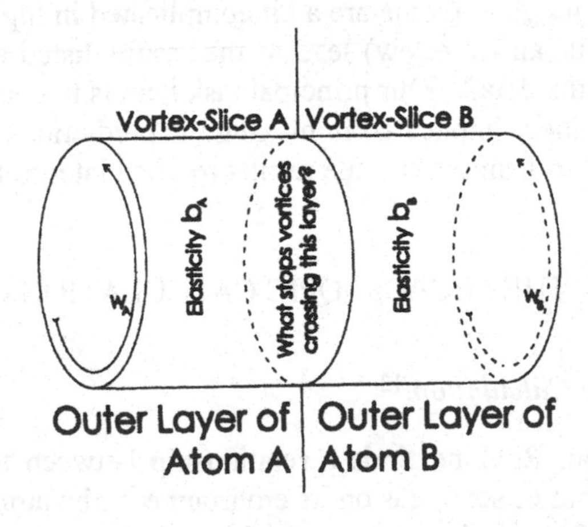

Figure 1

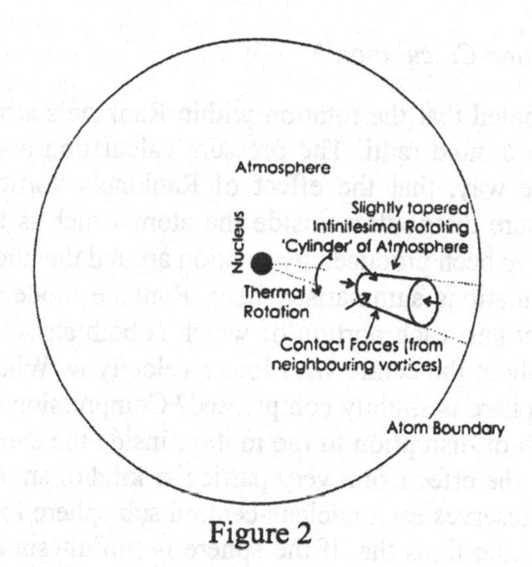

Figure 2

The properties of the ξ-component here are important to one of Rankine's successes, so their source needs to be considered. They are derived from stipulations about the attraction exerted by the nucleus on the atmosphere. Rankine assumes in particular that such forces vanish at the atomic boundary, then uses this assumption to perform a major re-organisation of an initial, superficially different, result (eq. 8, Rankine 1881, p. 26). So the ξ is no simple add-on term, incorporated after the bulk of the calculation is complete: it emerges from analysis of the complex interaction between attractions and vortical effects.

4.2. *The Temperature Calculation*[15]

The next of Rankine's results is that which extracts a link between microscopic parameters and macroscopic temperature. Rankine determines the conditions that ensure equilibrium when two alien vortices meet end to end (as in fig. 2). They can be thought of as vortices in the outermost layers of two touching atoms, one of chemical element A and the other of element B. He finds that there will be no tendency for one vortex to expand into the other if $w_A^2/b_A = w_B^2/b_B$ (with subscripts used to refer to the two different elements). This in turn suggests there is no tendency for heat to flow from one atom to the other, and that means (almost by definition) that they are 'of equal temperature'. Temperature, in other words, is determined by w^2/b; so, for each temperature scale τ, there must be a function Φ_τ, such that $\tau = \Phi_\tau(w^2/b)$.

4.3. *The Interaction Calculation*[16]

I have already noted that the rotation within Rankine's atom is not around the nucleus, but around radii. The pressure calculation above turns out to show, along the way, that the effect of Rankine's vortical motion is to produce a pressure distribution inside the atom which is the same as that which would have been produced by rotation around the nucleus.[17] So in the last of the calculations summarised here, Rankine models his atom by a sphere of perfect gas, each portion of which is both attracted to the centre, and in motion about the centre with linear velocity w. What happens to the rotation if the sphere is slightly compressed? Compression could, of course, produce all sorts of disruption to the motion inside the sphere, but Rankine only asks about the effects of a very particular kind of smooth compression — one which preserves each nucleus-centred sub-sphere intact. Under such conditions, Rankine finds that if the sphere is 'infinitesimally' compressed so that its average density increases slightly, from D to D+dD, then the increase in kinetic energy[18] can be especially simply expressed. For there is a function of D and w, viz. $\underline{\underline{\mathrm{F}}}(D,w)$, whose associated infinitesimal variation d$\underline{\underline{\mathrm{F}}}$ is very simply related to the increase in kinetic energy. For (if the mass of the gas is m) the latter is just $1/3mw^2.(\mathrm{d}\underline{\underline{\mathrm{F}}} + \mathrm{d}D/D)$. The most important thing here is that $\underline{\underline{\mathrm{F}}}(D,w)$ is just a function of D and w — for it this fact that will later yield the state function that can be identified with the entropy.

5. MACROSCOPIC EXPLANATIONS AND PREDICTIONS

Such then are Rankine's microscopic calculations. We can now start assembling the important macroscopic results he derives from them, focussing on those that yield empirical support for his theory in general, and aiming especially at the novel predictions promised in the title of this essay. Our scrutiny will not however restrict itself to predictions alone, for it is important to include explanations offered by Rankine for effects that were already known. Firstly, because there is no natural disentangling of the two — explanations and predictions — in the elaboration of Rankine's calculations; and secondly, because there is no agreement among observers of science that novel predictions deserve to be given any special epistemic weight.[19]

Some philosophies place special emphasis on the epistemic function of predictions because it is sometimes easy to bring a theory into conformity with a group of facts, which are already familiar. Knowingly or unknowingly, scientists can steer a 'derivation'. But with prediction, especially of novel facts, such distortion seems impossible. But this is not completely true, for knowledge that a novel effect has been predicted may well affect an experimenter's interpretation of the data that is alleged to confirm the effect's existence. In reviewing the empirical success of Rankine's theory, we will simply need to be sensitive to such matters.

5.1 First Steps: Perfect Gasses

From the 'pressure calculation' above, Rankine quickly derives the relationship between the pressure P exerted by a gas, its density D, and the speed w of vortical rotation, for he assumes there is no empty space in a gas — so the pressure it exerts on its container is simply that of the atmospheric fluid at the boundary of the atom, adjusted for any attraction exerted by the nuclei on the container. i.e. eq. 1 immediately yields (Rankine 1881, pp. 27, 238):

$$P = bD.(1+w^2/3b).(1-\xi(D,w)) + \zeta(D).$$

[equation 3]

where $\zeta(D)$ (representing the action of the nuclei on the container) becomes small as D becomes small, while ξ behaves as specified earlier.

In the opinion of Rankine (e.g. Rankine 1881, pp. 20-1) and his contemporaries (cf. Hutchison 1976, *passim*)), the especially simple laboratory behaviour of rarified hot gases resulted from the fact that action-at-a-distance forces within such gases were relatively small, and these in turn were thought of as approximating some 'ideal' or 'perfect' gas, in which such forces were zero. So Rankine's result shows (Rankine 1881, p. 28) that for such a material:

$$P = bD.(1+w^2/3b)$$

[equation 4]

Turning now to the temperature calculation, we note that if t is that temperature scale t for which $t \equiv \Phi_t(w^2/b) \equiv \kappa(1+w^2/3b)$ (where κ is a scaling factor that allows fixed points to be conveniently chosen), then a perfect gas has the familiar equation of state $P \propto t.D.$[20] So such t-scales are those provided by perfect-gas thermometers, and Rankine now has a vital link between routine laboratory temperature measurements, and the thermal parameters of his microscopic models.

Note that this κ is the temperature at which his thermal motions cease, i.e. the value of $\Phi_t(w^2/b)$ when $w = 0$. Gas-temperature is not then zero, because Rankine's theory allocates a small residual pressure to all substances, even in the absence of thermal motion — that *Ur*-elasticity resident in the stationary atmospheres. But κ ends up being a small number (2° or so on the modern Kelvin scale), so Rankine sometimes ignores it. Sometimes also, he uses a slightly different scale T ($\equiv t\text{-}\kappa = \kappa w^2/3b$),whose zero does indeed mark a complete absence of thermal energy. He argues that this latter scale satisfies what later becomes the standard criterion for 'absolute' temperature,[21] so it ends up as Rankine's preferred scale.

With this T-scale before our eyes, a small aside (to introduce some ideas used further below) is appropriate, before returning (in the next paragraph) to equations of state. Since the heat Q contained in some object is (for Rankine) just the kinetic energy of its vortical motions, Q must be directly proportional to this T. So for each object, there will be some constant K that makes $Q = K(t\text{-}\kappa) = KT$, a result (subsidiary to the temperature calculation) used in some of Rankine's arguments. Rankine calls this constant the 'real' specific heat of the object, to distinguish it from the various 'apparent' specific heats possessed by objects. The latter refer to the way the object interacts with its environment, for they refer to the heats *absorbed* in various changes, and depend on the characters of the changes: the former refers only to what happens inside the object. Its constancy is in part a matter of convention, for its properties depend on the temperature scale chosen; but the fact that it is constant on any scale linearly related to the perfect-gas scale is more substantial — this is a consequence of Rankine's model, via eq. 4. In a perfect gas (as we later see) it equals the apparent specific heat at constant volume; and since Rankine suspected *all* materials could found in the perfectly gaseous form, it followed that the parameter was empirically accessible — albeit often unknown.[22]

We have just seen that the temperature calculation, enables Rankine to derive the equation of state for a perfect gas from the pressure calculation. In doing this he also gets the Boyle-Mariotte law, of course, but Rankine never seems to have claimed support from the fact that his theory could explain this phenomenon. For kinetic explanations of this result were quite old by Rankine's time.[23] It seems indeed that all sorts of microscopic models are capable of giving this simple result; so though it may act as gatekeeper, it lacks the ability to distinguish one plausible kinetic theory from another. In Rankine's case, the situation is a little worse, for (as we have seen) the result is built into his theory from the start: for his atomic atmospheres already obey the Boyle-Mariotte law when stationary. So it is no surprise that they do the same when moving.

Rankine's derivation of the full perfect-gas equation $P \propto t.D$ is far more substantial a result, but again Rankine appears to have felt no urge to use this fact to support his theory. And rightly so. For the equation was quite familiar in Rankine's time, and one cannot feel confident that the temperature result was not guided by Rankine's knowledge that he needed to end up with the perfect gas equation.

Furthermore, the steps that develop this equation out of the Boyle-Mariotte law involve a confusing mixture of fact and convention, for the perfect gas equation only holds for appropriately chosen temperature scales, so its extra empirical content is hard to pinpoint. Hard facts are certainly involved, but they do not make any guns smoke. Rankine's own discussion clearly reflects the tension between convention and fact, and he argues, for instance, that what his calculation shows is that gas thermometers provide 'the most *convenient* measure of temperature' (Rankine 1881, p. 29, my emphasis).

Yet elsewhere, he does claim credit for a theoretical prediction that is highly dependent on this very convention — the constancy of the specific heats of a perfect gas (a 'truth' which is highly sensitive to the temperature scale chosen).[24] In this case, the facts were very much up in the air when Rankine wrote, so the circumstances are not analogous to those surrounding the equation of state. When the data turned up a few years later, Rankine's prediction was verified. His argument will be scrutinised below.

5.2 First Prediction: Equations of State for Imperfect Gasses

But when the pressure and temperature calculations are applied to fluids which are not perfect, the situation is quite different. Rankine developed his theory in the early 1840s, but set it aside for lack of empirical data. A few years later, the requisite data had been assembled, thanks to Regnault's systematic explorations of gaseous properties, and Rankine then found that Regnault's data was remarkably concordant with his own theory. In particular, he observed that Regnault's measurements of the saturated vapour pressure of steam, agreed with an equation derived via his theory, and did so extraordinarily well — 'as close[ly] as the precision of the experiments render[ed] possible' he claims (Rankine 1881, p. 38), 'through a range of temperatures from 30° below zero... to 230° above it, and of pressures from 1/2200 of an atmosphere to 28 atmospheres'. So Rankine rushed the formula into print, then prepared an account of the theory that lay behind it, revealed the following year.[25] In this case, the theory had been developed in complete independence of the empirical data, and Rankine was clearly right to see the data as providing significant evidence, even though no concrete prediction of an equation of state was involved: the theory only indicated a general form for this equation. And even that had not been published until the data confirmed it had appeared. So it is only a prediction in a relatively limited form.

His explanations start from equations 1, 2 (or 3) above. For when combined with Rankine's *t-w* linkage, these suggest real gases might have equations of state of the form

$$P \propto D.(t - B(D)/t) + A(D)$$

[equation 5]

where $A(D)$ and $B(D)$ are functions associated with the gas. Rankine attempted to fit Regnault's data to such an equation, and found that the behaviour of carbon dioxide (for instance) could be well represented by taking $A(D) \propto D^2$ and $B(D) \propto D$, which gave an approximate equation of state of the form[26]

$$P = \alpha\, tD + \beta D^2/t + \gamma D^2,$$

[equation 6]

where α, β and γ are constants, selected to fit the data. It is this equation that serves to provide the 'next' of Rankine's 'predictions' in our listing above — that for the cooling effect in the Joule-Thomson experiment.

To ensure that his method did not somehow beg the question here, Rankine systematically avoided using some of Regnault's results in determining these constants. This way he could employ the unused data 'as a test of the soundness of [his] theoretical reasoning'. Thus, using some measurements made at constant *volume*, and others at constant *temperature*, he 'predicts' the mean expansion of gases at constant *pressure* over the range 0-100°C. Comparing his results with Regnault's data (accurate to 0.0000136, according to Regnault) he finds the following:[27]

Gas	*Pressure*	*Predicted Value*	*Observed*	*Error*
CO_2	1.000	.0036988	.0037099 .003719	.0000111 .0000202
CO_2	3.316	.0038430	.0038450	.0000020
Air	1.0000	.0036650	.0036706 .003663 .003667	.0000056 .0000020 .0000020
Air	3.224	.0036955	.0036944	.0000011
Air	3.4474	.0036969	.0036965	.0000004
H_2	1.000	.0036598	.0036613	.0000015

Similar reasoning, supplemented by estimates of the interaction between the atoms of a liquid and those of its vapour, when the two are in equilibrium led Rankine (1881, pp. 40-3) to suspect that the saturated vapour pressure P of a vapour was related to (perfect-gas scale) temperature t, via an *approximate* equation of the form $log\ P = \delta + \varepsilon/t$ (where δ and ε are constants). But Regnault had already shown such a formula to be inadequate, so Rankine asks himself how the approximation might be improved, and guesses that he needs to add a $1/t^2$ term, to yield the modified relationship (Rankine 1881, pp. 43-4),

$$Log\ P = \delta + \varepsilon,/t + \eta/t^2$$

[equation 7]

where η is a further constant. It was this equation that gave the astonishing agreement with Regnault's data (Rankine 1881, pp. 44-6). It is also essential to another of Rankine's predictions — that for the specific heat of saturated steam.

5.3 Second Prediction: The Joule-Thomson Effect

Using one of these successful equations of state, Rankine and Thomson were later able to calculate theoretically the magnitude of the cooling effect suffered by carbon dioxide in one series of the Joule-Thomson experiments (though there is some doubt whether the calculation was done before or after the empirical data was in). These experiments were systematically carried out in the first half of the 1850s to solve one of the critical riddles of early thermodynamics — how accurate was 'Mayer's hypothesis'? This question turns up in the discipline in a variety of more or less equivalent forms, but amounts to asking how far *real* gas thermometers depart from *ideal* absolute scale — the 'hypothesis' declaring the two identical. This question is answered by using the experiments to prepare the translation tables which link the various gas scales, one by one, to the absolute scale.[28] A routine calculation enables these tables to be determined from measurements of the drop in temperature, as a gas is steadily pumped through a porous barrier. The experiment's virtue is it great sensitivity, and because of this, the calculations have to be especially accurate, and in particular, to accommodate deviations from the perfect gas equation of state.

Rankine (as we have seen) believed that the ideal temperature scale T (the one whose $\Phi_T(w^2/b)$ was proportional to w^2) differed from the perfect gas scale by the constant κ introduced above, so to him the experiment functioned to determine the value of κ. To us today his κ is a completely spurious quantity, with no reality whatsoever. In Rankine's time, it was certainly a dubious parameter, dependent on a highly speculative theory of the constitution of matter. Its credibility would surely have been greatly *reduced* if there was no way of determining its value, and Rankine certainly met this obstacle successfully. But he did a lot more, for he determined the parameter from a variety of experimental results, and got remarkably consistent results (Rankine 1881, pp. 330, 376-7). His 10 values were: 1.83; 1.08; 1.51; 2.09; 2.087; 2.345; 1.683; 1.762; 2.228; 2.14 (all in what we call degrees Kelvin). To Rankine this spread of values was good evidence for his theory — and we must surely agree. Given that his theory is false, the success of these predictions remains something of a mystery. Analysis of the experiments using conventional thermodynamics can explain why his values are all small numbers (when compared with everyday temperatures). But why the spread of results is so consistent — there are no negative values, for instance — remains puzzling.

With κ determined however, the calculation — above — that leads from (cooling effect plus equation of state) to κ can be reversed. So with κ taken as zero, the equation of state could be used to predict the cooling effect. In a letter of 1854, Rankine sent Thomson such a calculation for carbon dioxide — a gas which deviates somewhat more than others from 'perfect' behaviour. Thomson used it to calculate temperature changes of 8.27°, 8.07°, 4.96° (in three instances of the experiments he carried out with Joule). The measured values turned out to be 8.33°, 7.89°, 4.78°. Rankine quite rightly regarded this as further evidence for his theory.[29]

Like this prediction of the cooling effect for carbon dioxide, Rankine's prediction that saturated steam has a negative specific heat makes vital use of one of the above equations of state, that for saturated vapour (eq. 7). But it is also a particular case of a far more general problem, 'Rankine's problem' I call it — that of calculating *any* specific heat, a problem that Rankine attacks head-on, via the 'interaction' calculation above. So before we can examine this prediction, we need to make a detour, to inspect Rankine's use of that calculation's result. The detour will be an elaborate one, for this calculation is the main step in the path that leads to the most startling of Rankine's predictions, the existence of entropy.

5.4 Third prediction: Entropy

Rankine was an engineer, and his thermodynamics was focussed on the problem just articulated, a question of immense practical importance: how do we estimate the heat required to effect a specified change of state on a specified system? The entropy function S provides a direct and general answer to this question — at least for reversible changes. For if T is absolute temperature, and S the entropy of the system, the heat required to effect some infinitesimal change is simply $T.dS$. So the heat ΔH absorbed in some finite change is just the line integral $\int T.dS$ (along the path representing the change).[30] Rankine explicitly introduced the entropy in 1854 (though he called it the 'thermodynamic function'), and its role in solving Rankine's problem was widely noted by contemporaries. This was some ten years before the famous discussion of the function published by Clausius in 1865 — when the principle of entropy *increase* was announced (and the function given its modern name).[31] Rankine by contrast had little interest in irreversible thermodynamics and more or less ignored the issue, though he did make a couple of half-hearted attempts to deny that dissipation really occurred. In one them (published shortly *after* Clausius' announcement of the principle of entropy increase), he went so far as to claim that entropy is *conserved*![32]

Although Rankine did not single out his 'thermodynamic function' until 1854, the solution that it provided to 'Rankine's problem' was developed a few years earlier, though the expression of the solution was then a little more complicated. But all the hard work had then been done, and the path onwards to the entropy was very short. The vital step is taken in the 'interaction calculation' above, and it is very simple indeed to derive the entropy from its result. All that is involved is a simple reorganisation of the result, though I have severe doubts about the validity of one step.[33]
For consider the (adiabatic[34]) compression of a sample of gas made up of Rankine atoms. Each atom will then be slightly compressed, and the heat within unit mass of the gas (just kinetic energy of its thermal motions)will (by the interaction calculation) increase by $1/3w^2.(d\underline{\underline{F}} + dD/D)$, or $bT.(d\underline{\underline{F}} + dD/D)/\kappa$, when expressed in terms of the temperature-scale $T \equiv \kappa\overline{\overline{w}}^2/3b$. Because of energy conservation, this is also $-dL$, if dL is used for the total work done during the compression against internal and external forces. So

$$dL = -bT(d\underline{\underline{F}} + dD/D)/\kappa$$
[equation 8]

or, since DV is constant (when V is the volume of the sample of gas),

$$dL = T.dF$$
[equation 9]

with $F(T,V)$ another state-function, just $b(\log V - \underline{\underline{F}})/\kappa$.

Rankine now seems to assume that the same is true of a change that is not adiabatic, and in such a change the heat dH absorbed by a system is just the sum of the heat converted into work and the heat that remains as heat —

kinetic energy — within the system. The latter is just $K.\mathrm{d}T$, where K is (as before) the real specific heat of the system — a constant. So $\mathrm{d}H = K.\mathrm{d}T + \mathrm{d}L = K.\mathrm{d}T + T.\mathrm{d}F$, or (with S as the state function $F + K.\log T$),

$$\mathrm{d}H = T.\mathrm{d}S.$$
[equation 10]

But this is the precise relationship that defines the entropy!

5.5. *Fourth Prediction: The Specific Heats of Perfect Gasses*

We have already observed that Rankine includes a calculation in his 1850 papers, which indicates that perfect gases will have constant specific heats, and later claims credit for making this prediction, when the result is confirmed. His 1850 argument (at Rankine, 1881, pp. 253-5) is a particularly simple application of the interaction calculation. For evaluating specific heats is a prime example of 'Rankine's problem' and the natural way to deal with it using the mature theory would be via the entropy — harnessing eq. 10.[35] But since Rankine reached his result in 1850, such systematic machinery was not available to him, and he had to proceed more clumsily. At the heart of his calculation, however, is what amounts to a calculation of the entropy of a perfect gas from its supposed internal microscopic structure. This is the only such calculation I can detect in Rankine's work, and it is feasible because in this special case, the internal forces are all zero. In consequence the F-function supplied by the interaction calculation (as in eq. 8) is just κ/t — a result that will also be used with steam below.

From this it follows directly that the specific heat C_V — just $(\mathrm{d}Q+\mathrm{d}L)/\mathrm{d}t$ — when a sample of perfect gas is heated at constant volume (from temperature t to $t+\mathrm{d}t$) is $K + b(\kappa/t - \kappa^2/t^2)$, where K is the real specific heat. Now in 1850 Rankine suspected κ to be small (though this was not confirmed for a few years), and everyday temperatures are quite high on the gas-scale, so Rankine was able to ignore the second term here. (It certainly becomes negligible if the temperature is high *enough*.) Hence his very general claim (noted earlier) that real specific heats were empirically accessible; and the more concrete claim that the principal specific heats of a perfect gas were constant. A similar calculation shows the same to be true of C_P — the specific when the gas is heated at constant pressure (Rankine 1881, p. 255). In Rankine's time this was a bold prediction, for in the first half of the nineteenth-century, it was widely believed that gaseous specific heats varied considerably. Such a variation had been confirmed experimentally; and Carnot had used his theory to impose law upon it.[36] But (as we have already observed), Regnault's authoritative data confirmed Rankine's conflicting prediction a couple of years later.[37]

5.6. Fifth Prediction: The Specific Heat of Saturated Steam[38]

The fact that saturated steam has a negative specific heat is predicted using a very similar calculation, just a little more elaborate. This result seem paradoxical at first, but the phenomenon is already quite familiar, for the Wilson cloud chamber depends on it: because water vapour expands adiabatically, cools and becomes unstably super-saturated, it condenses as soon as a perturbation occurs. To prevent this tendency to condense, to keep the vapour exactly saturated, heat would have to be supplied during the expansion — even though the temperature is falling. The specific heat, in other words, is negative.

Another way to think about it is to consider the reverse of this cloud-chamber process, the compression of some saturated water vapour ('steam'). Work is done on the steam, the temperature rises and heat is abstracted. In this compression, *so much* external work is done, that some heat generated from that work is left over, for release to the environment. So the key to the phenomenon is the large amount of external work done.

Rankine's approach is very similar, except that he believes he can calculate (from the real specific heat) the amount of the generated heat that remains in the steam as the temperature rises slightly (by $\mathrm{d}t$, say). He can also use eq. 8 above to calculate the total work done: for he assumes that steam too has negligible internal forces,[39] so concludes (exactly as in the perfect gas case above) that $\mathrm{d}F$ is negligible. And eq. 7 allows $\mathrm{d}D/D$ to be expressed in terms of t and $\overline{\overline{\mathrm{d}}}t$. Given all these ingredients it is easy to calculate the heat released to the environment at different temperatures — as a function of the empirical parameters of eq. 7. When the values are plugged in, the specific heat turns out to be negative.

6. ARE RANKINE'S PREDICTIONS REALLY COUNTEREXAMPLES?

Such then are Rankine's predictions. They have been outlined here because they seem to undermine an argument for realism, that which presumes successful predictions can only be generated by hypotheses that are true. More precisely put, my target is the thesis that the hypotheses which *genuinely contribute* to an argument that makes some predictions must be *approximately true*, if the predictions succeed — in some substantial and non-trivial manner. The thesis presumably allows that odd flukes will sometimes occur, one-off astrological successes, for instance, amid a sea of failures. So its focus must be on sustained success, but given that, it claims more than just the improbability of success.

This is a historical thesis, and Rankine's thermodynamics refutes it — though it does not, of course directly refute the milder claim, that such events are rare. What makes Rankine a refutation is the fact that the arguments supplying Rankine with his sustained empirical successes depend on claims about heat and matter which are not plausibly judged as approximately true. So let us now review the complicated analysis of Section 5, to check that this is a reasonable interpretation of the story told above.

It might seem that I need to possess two things if I wish to deny that Rankine's hypotheses are approximately true: a knowledge of what really is true; and a decent theory of verisimilitude. But that is not so. For the fact that Rankine's thermal hypotheses are dramatically different from modern views about the sources of thermal behaviour is very good evidence that Rankine's theories are not true, and is the normal standard adopted for such purposes.[40] Furthermore, we can also use modern views as the yardstick by which to decide if Rankine's theory is *approximately* true, by asking to what extent the former can accommodate the latter, and how much they differ.

When we do this we note that there is much overlap between the two theories: both accept Newtonian mechanics (or some approximation to it); both accept that heat is often caused by microscopic motions; both focus on kinetic energy; and both accept the conservation of energy etc. So Rankine's theory is *partly* true. But that is not the issue: all sorts of nonsense is partly true. What matters here is that other parts, important ones, of Rankine's hypothesis are badly wrong. This is so because the modern theory paints a picture of the motions *radically* different from that important to Rankine: thermal phenomena are now seen as produced by motions *of* the molecules, not motions *within* them; pressure is produced by sudden collisions, not steady centrifugal effects; and the associated irregularity in the motions generates important thermal phenomena.

Furthermore, Rankine is not just missing these important facts about heat, he is denying them.[41] Given this, it seems reasonable to refuse to allow that Rankine's theory is approximately true here. One can plausibly do this without presenting any specific account of verisimilitude, for if such an judgments were systematically impossible, there would be no means available to us to assess theories of approximate truth. The decision is not final, and a clever account of verisimilitude might inspire me to reverse it, but I stick to it until convinced otherwise.

So now we only have to confirm that the untrue portions of Rankine's hypothesis were 'genuinely involved' in making his predictions. Again I lack a formal theory which tells me how to decide what premises are genuinely involved in reaching a conclusion, but our intuitions seem reasonably reliable here.[42] So let us go back over Rankine's empirical successes, checking that the vortical motions (something seriously wrong by modern standards) plays a significant role in the argument that leads to the prediction (or explanation).

What we will conclude is no big surprise to anyone who has waded through the story above: all of Rankine's successes really do use the model of thermal motions sketched above. This fact can be roughly summarised in the following logic-diagram (where the arrows represent chains of argument, not valid deductions — for we have to accommodate blemishes in Rankine's analysis. Dotted links represent special problems elaborated below.):

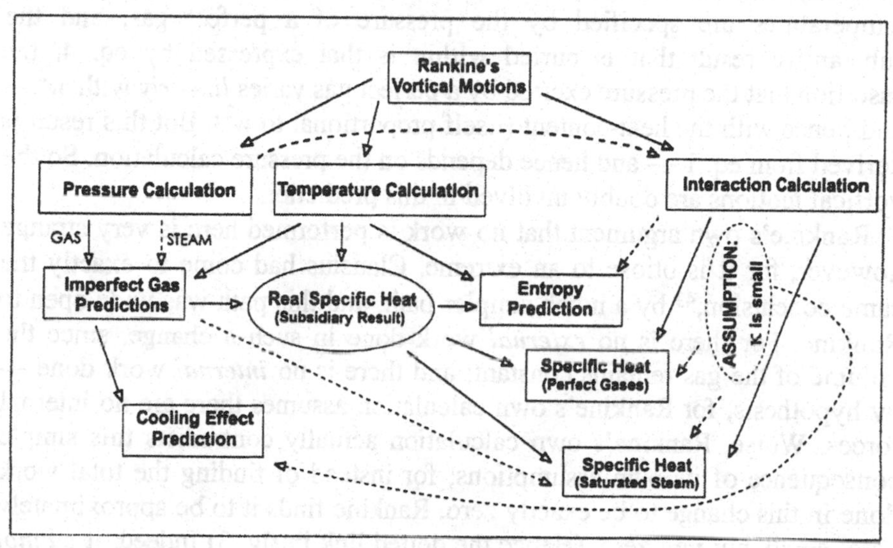

Figure 3

The first of the successes set out above, was that embodied in eq. 6, the approximate equation of state for a non-perfect gas. This in turn depended on the *t-w* linkage (i.e. the temperature calculation), which Rankine derived from a calculation that explored the stability of the contiguous vortex-slices depicted in fig. 2. So the internal motions are definitely involved in the temperature portion of this calculation. But they are also integral to the exact form of the approximation, for the powers of t and D that turn up in eq. 6 are derived from eq. 2, and that in turn depended on an analysis of the equilibrium of the tapered cylinder in fig.1. So again the thermal motions are deeply involved — via the two paths shown in fig. 3.

The successful prediction of the cooling effect for carbon dioxide depends directly on two things: the small magnitude of κ; and the equation of state. The former was not predicted by the theory, so its experimental confirmation does not support the theory. But the form of the equation of state was a theoretical prediction, so the experiment simply provides further evidence for Rankine's ability to explain equations of state. And that, we have just confirmed, did involve the false thermal motions.

Rankine's prediction that C_V is constant for a perfect gas depends on two subsidiary results: the constancy of the real specific heat K; and the fact that no work, internal or external, is done when a perfect gas is heated at constant volume. The latter of these claims was derived from the interaction calculation (so definitely depends on the vortical motions) plus the belief that κ is relatively small (a belief that does not reflect on the thermal motions). But the constancy of K is dependent on the convention that temperatures are specified by the pressure of a perfect gas, and the substantive result that is buried within is that expressed by eq. 4, the assertion that the pressure exerted by a perfect gas varies *linearly* with w^2 — and hence with the heat-content (itself proportional to w^2). But this result is derived from eq. 1 — and hence depends on the pressure calculation. So the vortical motions are doubly involved in this prediction.

Rankine's own argument that no work is performed here is very strange however, for it is otiose to an extreme. Clausius had come to exactly the same conclusion,[43] by a much simpler path, and this path was wide open to Rankine. For there is no *external* work done in such a change, since the volume of the gas remains constant; and there is no *internal* work done — by hypothesis, for Rankine's own calculation assumes there are no internal forces. Worse, Rankine's own calculation actually contradicts this simple consequence of its own assumptions, for instead of finding the total work done in this change to be exactly zero, Rankine finds it to be approximately zero: small, but non-zero. (Hence the dotted link in fig. 3) Indeed, it *cannot* be zero. For then κ becomes zero; yet if that happens b becomes zero too, and the whole theory collapses, for the atomic atmospheres lose their innate elasticity! It is quite unlikely that Rankine would have been put out by this discrepancy, as he never displays much interest in the niceties of argument, and both routes lead to the same practical result. In any case, the problematic result is announced at the very dawn of thermodynamics, when there was much confusion about what could be trusted. Thomson himself was suspicious of Clausius' argument, and was soon to be remarkably successful in developing a phenomenological thermodynamics, which systematically avoided any involvement with microscopic forces and motions etc.

Moving on, the entropy calculation is also dependent on the vortical motions. This too uses the *t-w* link derived from fig. 2, for it (firstly) replaces w^2 by the absolute temperature T (to get eq. 8); and (secondly) assumes the real specific heat constant as part of the process of simplifying the result (to get eq. 10), and that constancy followed from the temperature calculation. But the entropy argument also uses the interaction calculation, which in turn is utterly saturated with the internal motions of the Rankine atom (in part, via the pressure calculation — hence the dotted link). If such motions were to be ignored, the calculation would simply evaporate. But as indicated in my analysis,[44] I am not convinced that Rankine's successful argument here is actually valid, for he seems to generalise too hastily from the adiabatic result — hence the dotted line.

Something similar is true of the final prediction — that for saturated steam, but the situation is clouded by confusing complications. For the *first* of the relationships between steam pressure and *t* that Rankine derives from the pressure calculation is the one that fails the empirical test (hence the dotted links in fig. 3), yet the argument in favour of it is vital to the derivation of the second — successful — relationship: for that is generated by applying a routine mathematical correction (that has nothing to do with the theory) to the first relationship! I suspect that Rankine judged that the second relationship could be derived directly from the theory, via a more elaborate calculation, but saw no point in doing so. This assessment seems reasonable. But the fact remains that one step in the prediction process is simply a lucky guess — and this is precisely the issue here. 'Miracles' do occur.

There is another peculiar feature of this prediction worth observing here. Although the interaction calculation is genuinely involved in the argument that makes the prediction, the final result is peculiarly insensitive to it. And exactly the same is true of Rankine's other specific heat prediction, that for perfect gases. Its contribution to the predicted effect is *negligible*: exactly the same observable effect could be predicted by very different versions of the interaction calculation. So the pressure calculation is the only portion of Rankine's theory that I claim to be directly supported by the prediction of the constancy of specific heats. But that, of course, still involves portions of his theory that are badly wrong.

7. THE BIG MYSTERY: WHY DOES RANKINE'S THEORY WORK SO WELL?

The remarks of the above paragraph, give a small hint as to how Rankine's false theory can be so successful — for they indicate that although some of his hypotheses are genuinely involved in the *arguments* that lead to Rankine's conclusions, their impact on the *conclusion* is very small. In consequence, their falsity lacks the power to disrupt the conclusion too seriously. So when we reviewed Rankine's prediction that perfect gases have constant specific heats, we saw that the only feature of his theory that really influenced the conclusion was the fact that his theory required perfect-gas pressure to be a linear function of the kinetic energy of the thermal motions. On this quantitative point, his theory agrees with many other, more standard, versions of the kinetic theory, all equally capable of making the same prediction.[45] Rankine may well have made the prediction, simply because he was one of the first to reach a conclusion available to many: for although varieties of the kinetic theories had been around a long time before Rankine, sustained development of them was not a major scientific activity until after the acceptance of the first law — around 1850. So in this case, Rankine's success is quite compatible with 'structural' realism. His false theory generates a correct quantitative result, which then does all the hard work in yielding the prediction. And in this case too, the success is no huge mystery: the details of the kinetic model used do not seem to influence the prediction.

Much the same is true of Rankine's other specific heat prediction, that for saturated steam, for this too hinges on a quantitative intermediary — in this case, the P-t relationship for saturated steam (eq. 7). And although Rankine's used his hypotheses in the *argument* that yielded this amazingly accurate relationship, the role of the argument here was primarily heuristic. Rankine's method is very close to trial and error, and it is no mystery that this method can generate success (the issue is whether it is efficient!).

Rankine's explanation of the form of the equation of state for an imperfect gas, by contrast, arises directly from the pressure calculation, and there are no accurate quantitative intermediaries between the ontology and the prediction that might allow for a structural interpretation here. But in this case we can make rather different sense of Rankine's success by reviewing the range of possibilities open to him. If one knows that the equation of state for a perfect gas $P=DT$ is somewhat inaccurate when used for some *imperfect* gas, then standard techniques of approximation, make it highly likely that an equation of the form

$$P = DT + \alpha D^2T + \beta DT^2 + \gamma D^2T^2 + \delta D/T + \varepsilon T/D + \eta/TD$$

[equation 11]

(where the $\alpha, \beta, \gamma, \ldots$ are constants) will provide a better approximation (and so on with higher powers — and many more floating constants). So in the end, what Rankine's theory supplied is (perhaps) one of the steps in a predictable iterative procedure — available to those who did not share Rankine's ontology. But as the number of constants increases, the iteration would get very unwieldy. Rankine's hypotheses functioned to avoid such complications, by making many of the constants vanish. I can see no explanation for this efficiency.

The prediction of entropy is far more difficult to explain, and I can offer no genuine insight. If I am right that Rankine's derivation is invalid,[46] his argument is restricted to adiabatic changes. That given, his success here is a quite intelligible, but boils down to another lucky guess — the sort of thing the no-miracle realists tend to deny. For it is *trivially* true that there is a state-function S* that satisfies the relationship characterising entropy (viz., $\Delta H = \int T.dS^*$) for adiabatic changes. For in such changes, $\Delta H = 0$ (by definition[47]), so any *constant* S^* mimics the entropy! Rankine's achievement here is the declaration that this same relationship holds for non-adiabatic changes. While this further step does not (to me) seem justified, the fact remains that he did take this step!

It is perhaps worth adding, that Rankine does not get the numerical *value* of the entropy right either. He predicts that the function S exists (i.e. that $\Delta S = \int dH/T$ is independent of the path of integration), but his theory does not correctly estimate the magnitudes of ΔS. In conventional thermodynamics, the entropy is indeterminate to the extent of a harmless constant, that automatically disappears in any evaluation of an entropy change ΔS. But Rankine's entropy is far more indeterminate, for it turns out to include an arbitrary function of temperature, that has the potential to disrupt most calculations using Rankine's official method for evaluating the entropy. This potential was never actualised however, for evaluations of the entropy were a rare event in the early years of the subject. The problem disappears in the special case of perfect gases; and Rankine avoids using his own method in the other important case that comes up, the evaluation of the entropy of steam in his 1859 *Manual of the steam engine*. Here he appears to use a more conventional method, one that depends on little more than the definition of entropy (eq. 10 above).[48]

8. WHAT DID RANKINE'S CONTEMPORARIES MAKE OF HIS SUCCESSES?

Such behaviour — avoiding the details of the theory, and focusing on the concrete result it led to — was common among contemporary observers of Rankine's work. A generation of engineers used the formulae set out in the numerous editions of Rankine's *Manual of the steam engine*, yet one cannot imagine more than a handful of them delving into the vortical dynamics, and even that estimate is probably far too optimistic.

This phenomenon is very clear among scientists too: they used Rankine's results, enlisted him among the supporters of the new view of heat, but ignored the theory. Thomson's discussion of Rankine's carbon dioxide calculation (Sectoin 5.3 above) illustrates this perfectly: it contains no reference to Rankine's special theory of matter.

Though precise motives here are hard to determine, this behaviour is no mystery. Rankine's theory was obscure and speculative, and did not (in particular) conform to the phenomenological standards set by Thomson's own work — standards that soon prevailed as the thermodynamic norm. So in reviewing Rankine's work in 1878, Maxwell complains that it is 'difficult... to attach any distinct meaning to the total actual heat of a body' (1890, v.2, p. 664). Rankine certainly recognised this reaction to his work, for shortly after the initial successes set out in the 1850 papers, he embarked on a pair of projects that reveal this concern to us.

The first generated a series of attempts to divest, as far as possible, the vortical theory of the particular details it attributed to its hypothetical microscopic motions, and strip the theory back to its bare minimum 'steady closed streams of continuous fluid' motion.[49] The second was an attempt to find completely new foundations for Rankine's results, phenomenological foundations that made no reference at all to microscopic causes. This second project was undoubtedly a failure. It never fully escaped appeal to the vortex theory — which remained essential, to link absolute temperature with heat-content (e.g. Rankine 1881, pp. 321, 376.). And its axioms were inadequate[50] — either ill-justified rules-of-calculation (as in this 1867 version of the second law):

> To find the whole work... multiply the absolute temperature at which the change... takes place by the rate per degree at which the external work is varied by a small variation of the temperature.

Or frankly unintelligible (as in this 1857 version):

> If the absolute temperature of any uniformly hot substance be divided into any number of equal parts, the effects of those parts in causing work to be performed are equal.

Any 'student who thinks he can form any idea of the meaning of' such versions of the Second Law is capable of explaining anything, opined Maxwell (1890, v.2, p. 664)!

University of Melbourne

REFERENCES

Brush, S.: 1976, *The kind of motion we call heat*, v.1, North-Holland, Amsterdam.

Cardwell, D.S.L.: 1971, *From Watt to Clausius: the rise of thermodynamics in the early industrial age*, Heinemann, London.

Cardwell, D.S.L. and Hills, R.: 1976, 'Thermodynamics and practical engineering in the nineteenth century', *Histo y of Technology* 1, 1-20.

Carnot, S.: 1986, *Reflexions on the motive power of fire: a critical edition with the surviving manuscripts*, ed. & trans. R. Fox, Manchester University Press, Manchester.

Carnot, S.: Émile Clapeyron & Rudolf Clausius: 1960, *Reflections on the motive power of fire; and other papers on the second law of thermodynamics*, ed. E. Mendoza, Dover, New York.

Carnot, S.: 1824, *Réflexions sur la puissance motrice du feu et sur les machine propres à développer cette puissance*, Bachelier, Paris. Trans R. H. Thurston, ed. E. Mendoza in Carnot *et al.* (1960, pp. 1-59). Also trans. R. Fox in Carnot, 1986.

Clausius, R.: 1850, 'On the motive power of heat, and on the laws which can deduced from it for the theory of heat', trans. W. F. Magie, in Carnot *et al.* (1960, pp. 107-52). Originally: 'Ueber dic bewegende Kraft der Wärme, und die Gesetze, welche sich daraus für die Wärmelehre selbst ableiten lassen', *Annalen der Physik und Chemie*, ed. J. Poggendorff, 129(1850): 368-97, 500-24. Presented (but apparently not read) to Berlin Academy Feb., 1850.

Clausius, R.: 1865, 'On several convenient forms of the fundamental equations of the mechanical theory of heat' in Clausius (1867, pp. 326-65). English translation of 'Ueber verschiedene für die Anwendung bequeme Formen der Hauptgleichungen der mechanischen Wärmetheorie' [*Ann. Phys.*,201 (1865), 353-400, read 24 Apr 1865]. Translator not specified.

Clausius, R.: 1866, 'On the determination of the disgregation of a body, and on the true capacity for heat', *Phil. Mag.* (ser.3) 31, 28-33. Read 22 Aug 1865, and originally published in French in 1865.

Clausius, R.: 1867, *The mechanical theory of heat*, ed. J. Hirst, various translators (not specified), 1st ed, London. Second edition, 1879 is very different, and is no substitute.

Daub, E.: 1967, 'Atomism and thermodynamics', *Isis* 58, 293-303.

Daub, E.: 1970, 'Waterston, Rankine, and Clausius on the kinetic theory of gases', *Isis* 61, 105-6.

Fox, R.: 1971, *The Caloric theory of gases: From Lavoisier to Regnault*, Clarendon Press, Oxford.

Fox, R.: 1986, [Annotations etc.] in Carnot 1986.

Hutchison, K.: 1972, 'Der Urpsrung der Entropiefunktion bei Rankine and Clausius', *Annals of science* 30, 341-64.

Hutchison, K.: 1976, 'Mayer's hypothesis: a study of the early years of thermodynamics', *Centaurus* 20 279-304.

Hutchison, K.: 1976, *W.J.M. Rankine and the rise of thermodynamics*, D. Phil. thesis, Faculty of Modern History, University of Oxford.

Hutchison, K.: 1979, Review of Truesdell & Bharatha, 1977, *Annals of science* 36, 660-3.

Hutchison, K.: 1981a, 'W.J.M. Rankine and the rise of thermodynamics', *Brit. J. Hist. Sci.* 14, 1-26.

Hutchison, K.: 1981b, 'Rankine, atomic vortices, and the entropy function', *Archives internationales d'historie des sciences* 31, 72-134. Proof of this paper did not reach the author until *after* publication (!), so unfortunately there are numerous minor errors in the text.

Lyons, T.D.: 2002, 'Scientific realism and the pessimistic meta-modus tollens', this volume.

Maxwell, J.C.: 1890, *The scientific paper of James Clerk Maxwell*, ed. W.D. Niven, 2v (bound as one), Dover, New York, 1965. Facs. Reprint of 1890 original.

Pannekoek, A.: 1961, *A history of astronomy*, translator not identified, George Allen & Unwin, London.

PSA 1994: proceedings of the 1994 biennial meeting of the Philosophy of Science Association, 2 v, ed. D. Hull, M. Forbes & R. Burian, Philosophy of Science Association, East Lansing, MI, 1994.

Psillos, S.: 1999, *Scientific realism: how science tracks truth*, Routledge, London.

Rankine, W.J.M.: 1849, 'On an equation between the temperature and the maximum elasticity of steam and other vapours', *Edinb. New Phil. J.* 47, 28-42. The paper that was rushed into print to reveal the exceptional ability of Rankine's hypothesis to accommodate Regnault's data.

Rankine, W.J.M.: 1851, 'On the centrifugal theory of elasticity as applied to gases and vapours' *Phil. Mag.* (ser.4) 2, 509-42, reprinted Rankine, 1881, pp. 16-48. Read 4 Feb. 1850. The first of Rankine's two major 1850 expositions of his hypothesis, the one containing what I term the 'pressure' and 'temperature' calculations.

Rankine, W.J.M.: 1852, 'On the reconcentration of the mechanical energy of the universe', *Phil. Mag.* (ser.4) 5, 358-60. Reprinted (Rankine 1881), pp. 200-2. Read 2 Sept., 1852.

Rankine, W.J.M.: 1853, 'On the mechanical action of heat, especially in gases and vapours' [introduction, sections 1-IV and appendix to section IV], *Trans. R. Soc. Edinb.* 20, 147-90. Reprinted (Rankine 1881), pp. 234-84. Read 4 Feb. 1850. The second of Rankine's two major 1850 expositions of his hypothesis, the one containing what I term the 'interaction calculation'.

Rankine, W.J.M.: 1855, 'On the mechanical action of heat:- supplement to the first six sections, and section seventh' *Trans. R. Soc. Edinb.* 3, 287-92. Abstract only. Read 19 Feb. 1855.

Rankine, W.J.M.: 1857, ' Heat, Theory of the mechanical action of, or Thermo-dynamics', in of J.P. Nichol, *A cyclopaedia of the physical sciences*, 1st edn, London, 1857, pp. 338-54. Written 1855: see Tait (1881, p. xxii). A similar entry occurs on pp. 411-29 of the 2nd edn, published 1860.

Rankine,W.J.M.: 1859, *A manual of the steam engine and other prime movers*,1st edn, London. Many later editions until: 17th, 1908. Last prepared in Rankine's lifetime: 6th, 1873. I have not detected any differences in the thermodynamical section of the various editions.

Rankine, W.J.M.: 1865, 'On thermodynamic and metamorphic functions, disgregation, and real specific heat', *Phil. Mag.* (Ser.4) 30, 407-10. Dated 15 Nov 1865. A reply to Clausius, 1866.

Rankine, W.J.M.: 1881, *Miscellaneous scientific papers* ... ed. W. J. Millar; with 'A memoir of the author', by P.G. Tait, Griffin, London).

Siegel, D.: 1991, *Innovation in Maxwell's electromagnetic theory: molecular vortices, displacement current, and light*, Cambridge University Press, Cambridge.

Tait, P.G.: 1881, 'A memoir of the author [= Rankine]'. Dated 1880. In Rankine, 1881, pp. xix-xxxvi.

Talbot, G.R. & A.J. Pacey: 1966, 'Some early kinetic theories of gases: Herapath and his predecessors', *Brit. J. Hist. Sci.* 3, 133-49.

Thomson, W.: 1882, *Mathematical and physical papers*, v.1, University Press, Cambridge.

Truesdell, C. & S. Bharatha: 1977, *The concepts and logic of classical thermodynamics as a theory of heat engines rigorously constructed upon the foundations laid by S. Carnot and F. Reech*, Springer-Verlag, New York, Heidelberg and Berlin.

NOTES

[1]For a survey (and endorsement) of some versions of this argument, see Psillos (1999, pp. 70-97).

[2]For discussions of the argument see Psillos (1999, pp. 102-14); Lyons, (this volume).

[3]For a sustained development of such objections to the pessimistic induction, see Psillos (1999, pp. 115-45).

[4]For a survey of the historical context of my discussion, see Cardwell, (1971).

[5]Carnot makes a number of confirmed predictions using a theory that is now universally deemed to be false. However, it is not universally agreed that his false assumptions are essential to his reasoning processes, and Carnot (accordingly) has sometimes been interpreted along the lines sketched above for Ptolemy. While I have argued against this view elsewhere, until it is agreed that the conservation of caloric is vital to Carnot's arguments, a further separate case is worth making. So the Rankine example is not unique — just illuminating. (For the claim that Carnot does not use the conservation of caloric in his reasoning, see: Truesdell & Bharatha (1977, pp. 116, 118); Psillos (1999, p. 124). However, see Hutchison (1979, p. 662, esp. n. 2.).

[6] For details, see Hutchison (1976, pp. 291-2), analysing Clausius (1850, pp. 122-7, 135-6, 149). On pp. 126-7 here, Clausius explains why the Carnot theory requires a positive specific heat.

[7]See Hutchison (1976, pp. 293-4); Cardwell and Hills (1976, pp. 10-1).

[8]The bulk of Rankine's thermodynamical work is assembled in the posthumous *Miscellaneous scientific papers* of 1881, but two important discussions are not included there: Rankine, 1857 = *Heat, theory of.* (written in 1855); and the long-discussion in Rankine 1859 = *Manual of the steam engine* (1st edn, 1859). Virtually all of the present discussion of Rankine's investigations comes out of my 1976 Oxford doctoral thesis, *W.J.M. Rankine and the rise of thermodynamics*, but some of the results in that thesis have been published separately, notably in my: 'Der Ursprung der Entropie funktion...' (1973); 'Mayer's hypothesis...' (1976);'W.J.M. Rankine and the rise of thermodynamics' (1981); 'Rankine, atomic vortices, and the Entropy function' (1981). Daub has also published important studies of Rankine's thermodynamics, notably 'Atomism and thermodynamics' (1967).

[9]See Rankine (1881, pp. 21-4, 236-40).

[10]The contrast between the two theories is perhaps worth noting here, for the sake of clarity. Both theories *derive* a relationship between kinetic energy and temperature. Modern theories avoid talking about the heat 'inside' objects — which Rankine identifies, *ab initio*, with (some) internal kinetic energy.

[11]See note 37.

[12]I do not know when this prediction is explicitly confirmed, but the result was soon accepted. After meeting the prediction in Rankine's work, Thomson cites empirical evidence in its favour in a letter to Joule of 1850: see Hutchison (1976, pp. 293-4).

[13]This calculation amalgamates calculations in Hutchison (1981b, pp. 76-8, 80-7, 93-6).

[14]Cf.: Hutchison (1981b, p. 87, eq. 53); Rankine (1881, p. 27, esp. eq. 11).

[15]Cf. Rankine (1881, pp. 27-31, 236-7); Hutchison (1981b, pp. 96-100).

[16]This section is a rather free paraphrase of Rankine's second major 1850 paper, 'On the mechanical action of heat... sections 1-IV' = Rankine, 1853 reprinted Rankine (1881, pp.

234-84). For my final result, cf. Rankine's eq. 6, (1881, p. 249). For a more extended exposition, see Hutchison (1981b, pp. 88-92).

[17]Cf.: Hutchison (1981b, pp. 88, 101); Rankine (1881, pp. 237, 247).

[18]This terminology is anachronistic, but no damage is done. Rankine is one of the key figures in drawing the distinction between potential and kinetic energy, though he refers to the latter via the neo-Aristotelian term 'actual' energy. It is not identical to kinetic energy, but it is very close. Cf.: Hutchison (1981a, pp. 15-6).

[19]See, e.g., *PSA 1994*, v.2, pp. 133-64.

[20]Cf.: Rankine (1881, pp. 28-9) for a verbal paraphrase of this equation; p. 30 (eq. 16) for the t-w relationship involved.

[21]Because absolute thermometry had earlier been based upon the Carnot theory, this criterion was very uncertain for a few years: see Hutchison (1976, pp. 283-5, 299-300). For Rankine's argument that T meets it, see Hutchison (1981a, p. 10).

[22]See, e.g., Rankine (1881, pp. 21-4, 30-1, 239-40). I discuss these ideas on in Hutchison (1981b, pp. 76-7), noting that Rankine sometimes (inconsistently and problematically) allows K to vary, perhaps across a phase-change. In the core of his theory, this disruptive possibility is ignored.

[23]Brush (1976, p. 20) traces kinetic theories back to 1738 (D. Bernouilli), noting that Newton had included a static explanation of the Boyle-Mariotte law in the *Principia*. See also Talbot & Pacey (1966, *passim*).

[24]Rankine claims this credit in note A on p. 336 of (1881) and also on p. 377. When the predictions are made however (in Rankine 1881, pp. 255, eq. 20, 323-6 = §58 — perhaps also at pp. 257-8?), they are not singled out for special attention, but buried in the detail, with Rankine giving no hint they are some sort of test of his theory. As the predictions conflicted with prevailing opinion at the time they were made, Rankine may well have been attempting to de-emphasise an apparent defect in his theory.

[25]For the result, see Rankine, 1849 = Rankine (1881, pp. 1-12). For the account of the theory, see: Rankine, 1851 = Rankine (1881, pp. 16-48); Rankine, 1853 = Rankine (1881, pp. 147-90). These papers were read in Feb. 1850, and nominally published in 1851 and 1853 respectively.

[26]Cf. Rankine (1881, pp. 32-9, esp. eq. 21, p. 32). For the evaluation of $B(D)$ see Rankine's comment immediately following this equation; for the evaluation of $A(D)$ see eq. 28, p. 37. I single out carbon dioxide here, simply because we need the result further below.

[27]The quotation is from Rankine (1881, p. 38), the table from p. 39. For the data used to determine the constants, see p. 33.

[28]See Hutchison (1976, *passim*).

[29]See: extract from Rankine's letter to Thomson of 9 May 1854, reprinted on pp. 375-6 of Thomson, 1882. For the comparison between prediction and measurement, see: Thomson & Joule at Thomson (1882, pp. 375, 430); Rankine (1855, pp. 289-90).

[30]For Rankine's articulation of this question, see: Rankine (1881, pp. 343, prop. 1). For Thomson's version, see Thomson (1882, pp. 298-9). For Rankine's use of the entropy to answer it, see: Rankine (1881, p. 352, eq. 13); (1855, pp. 288-9).

[31]Clausius (1865, pp. 364-5). For Rankine's 1854 introduction of the function, see Rankine (1881, pp. 351-2).

[32]Rankine, 1881, p. 233 (from 1867). For a much earlier denial of dissipation, see Rankine, 1852. For a rare (but fleeting) acknowledgment of irreversible effects, see Rankine (1881, pp. 226-7, dating from 1855).

[33]See note 34.

[34]An 'adiabatic' change is one in which there is no exchange of heat between the system under consideration and its environment. Adiabaticity appears to be vital to the logic of the argument here (in the switch from kinetic energy to work that yields eq.8), but Rankine does not explicitly restrict himself to such a change. Yet if this restriction is added, it is not clear that the truth of eq.9 in the adiabatic case, enables us to conclude that it is also true in general. Rankine's logic here is very blurred, probably defective, and the equivocation may be essential to reaching the conclusion. For further discussion, see Hutchison (1981b, p.102, esp. n.75). My judgment here (and elsewhere) might seem to clash with that of Tait (1881, p. xxix) who endorses a suggestion by Maxwell (1890,v.2, p.662) that Rankine's logic is faultless. Tait however is a notoriously unreliable witness; and I take Maxwell to mean only that Rankine's logic comes up to the standards of the physics community (which rarely insists on really tight argument). Maxwell is also being diplomatic here (even though Rankine had died several years earlier), for it is also clear from the same source that Maxwell found many difficulties in Rankine. In particular, Maxwell criticises Rankine's theory for its use of 'heat-content', a notion which Maxwell declares to be unintelligible (p.664). Yet a couple of pages later, Maxwell changes tack, admitting the notion does make sense, but within kinetic theory (p.665). I do not regard his comments as reliable evidence against my view.

[35]This might seem to be begging the question, since entropy is often derived (in conventional thermodynamics) from the specific heats. But due to a subtle mathematical error, Rankine overestimates the power of his theory, and believes he can ascertain his thermodynamic function phenomenologically without the need for so much data. See Hutchison, (1981a, p. 17).

[36]For the earlier belief, see: Carnot in Carnot *et al.* (1960, pp. 29-31); Fox (1986, pp. 135-8, nn. 50, 59); Cardwell (1971, p. 213). For Regnault's refutation of it, see: Fox (1971, p. 298); Rankine (1881, p. 336, note A).

[37]According to Rankine's first biographer (Tait, 1881, p. xxix) Rankine also uses his early theory to predict, successfully, the actual magnitude of the specific heat of air.

38The calculation here is articulated more fully in Hutchison (1981b, p. 106)

[39]Rankine is in effect assuming that steam can be treated as a perfect gas. Thomson does the same, but Clausius does not. In the end, the assumption is deemed unreasonable: see Hutchison (1976, pp. 288-90, 296, 299).

[40]The argumentation here is rather indirect, and probably not worth articulating fully. For the falsity of Rankine's theory is beyond doubt, whatever the process that leads us to acknowledge this. But see Lyons (this volume).

[41]Rankine does in fact seem to allow erratic motions, but this does not detract from my case, because the irregularity he accommodates has no thermal consequences.

[42]Psillos formulates such a criterion (1999, p. 110), but (as Lyons has pointed out to me (and mentions in this volume), no appeal is made to this criterion in judging concrete cases (in Psillos' following chapter, pp. 115-45). Furthermore, Psillos suggests (1999, pp. 112, 128-9) that scientists are rather good at judging which components of a theory are tested by an experiment, so he clearly takes the view that informal procedures are more or less trustworthy.

[43]Clausius (1850, pp. 129-30).
[44]Especially note 34.
[45]See Brush (1976, pp. 166, 172-3). Rankine himself effectively shows that conventional kinetic theory leads to the same result, though it gets the actual value of the specific heat wrong: see Rankine (1881, p. 321-3) and Daub, 1970. For an exception (provided by Herapath, whose temperature was related to momentum), see Brush (1976, p. 111).
[46]See note 34.
[47]See again note 34.
[48]For the 1859 calculation, see Rankine (1859, p. 383-4). In reply to criticism from Clausius (1866), Rankine eventually admits that his 'official' method of evaluating entropy (etc.) only works for substances which become perfectly gaseous at sufficiently high temperatures: see Rankine (1865, p. 409).
[49]Maxwell (1890, v.2, p. 663); Hutchison (1981a, p. 22, n.19). Rankine long opposed the conventional kinetic theory because of its notorious incompatibility with measured specific heats: see, e.g., Rankine (1881, p. 323).
[50]The versions of the Second Law quoted below are from Rankine (1881, p. 434) and Rankine (1857, p. 341). For a systematic survey of Rankine's versions of the Second Law, see Hutchison (1976, pp. 250-66).

MICHEL GHINS

PUTNAM'S NO-MIRACLE ARGUMENT: A CRITIQUE

1. PUTNAM'S VERSION OF THE NO-MIRACLE ARGUMENT

More than a quarter of a century has elapsed since Hilary Putnam first proposed his famous 'no-miracle' argument in 'What is mathematical truth?' (1975). The argument, also known as the 'ultimate argument' (van Fraassen 1980), is still widely discussed and is considered by many to be, just as Putnam believed, *the* major argument in favour of scientific realism (Leplin 1997, Psillos 1999). In this paper, I will review various forms of the argument and show them to be unconvincing from a naturalist's point of view. My main point is that scientific realism is indefensible when it is conceived to be a scientific explanation of the success of science. This is not because other —'better'— explanations are available or because the no-miracle argument is logically flawed or because examples of false empirically successful theories can be produced. It is rather because scientific realism is a *philosophical* position and cannot possibly count as a *scientific* explanation of *any* fact. Essentially, I want to question the scientific explanatory force of truth and, at the same time, argue against a form of naturalism which construes scientific realism as a scientific hypothesis.

In 1976, in his John Locke Lectures, Putnam offers the following formulation of the no-miracle argument (NMA): 'Whatever the realists say, they typically say that they believe in a 'correspondence theory of truth':

> When they argue *for* their position, realists typically argue *against* some version of idealism — in our time, this would be positivism or operationalism. (…) And the typical realist argument against idealism is that it makes the success of science a *miracle* (…) The modern positivist has to leave it without explanation (the realist charges) that 'electron calculi' and 'space-time calculi' and 'DNA calculi' correctly predict observable phenomena if, in reality, there are no electrons, no curved space-time, and no DNA molecules. If there are such things, then a natural explanation of the success of theories is that they are *partially true* accounts of how they behave. And a natural account of the way scientific theories succeed each other (…) is that a partially correct/incorrect account of

S. Clarke and T.D. Lyons (eds.), Recent Themes in the Philosophy of Science, 121–137.

> a theoretical object (...) is replaced by a *better* account of the same object or objects. But if those objects don't really exist at all, then it is a *miracle* that a theory which speaks of gravitational action at a distance successfully predicts phenomena; it is a *miracle* that a theory which speaks of curved space-time successfully predicts phenomena; and the fact that the laws of the former theory are derivable 'in the limit' from the laws of the latter theory has no methodological significance.
> (...) That science succeeds in making many true predictions, devising better ways of controlling nature, etc., is an undoubted empirical fact. If realism is an *explanation* of this fact, realism must itself be an over-arching scientific *hypothesis*. And realists have often embraced that idea, and proclaimed that realism *is* an empirical hypothesis. But then it is left obscure what realism has to do with theory of truth. (Putnam 1978, p. 18).

This, admittedly lengthy, quotation shows how complex the matter is. The nature of reference, truth and explanation, scientific progress, the role of methodology and the epistemic status of scientific realism all crop up in the context of the no-miracle argument. For the sake of clarity, an attempt at reconstructing a simplified version of the argument, or better, arguments in the plural, is in order.

2. A RECONSTRUCTION OF PUTNAM'S THREE ARGUMENTS

The success of science is supposed to be an empirically ascertainable fact which calls for an explanation. Putnam argues[1] that scientific realism (SR) — construed as the thesis that mature theories are partially true and that their theoretical terms have real referents — provides the only plausible explanation of this fact. Since the global success of science is parasitic on the success of individual theories, it is wise to concentrate first on a particular theory before discussing the broad issue of the success of the scientific enterprise in general. Putnam in fact makes a distinction between three kinds of success which we may call *predictive, progressive* and *methodological* and this leads to three distinct arguments in favour of the truth of scientific theories.

For a given scientific theory T it must be possible to determine if it accurately predicts the phenomena within its domain (if not, T would presumably cease to qualify as scientific). If it does, the theory under examination can be said to be *predictively* successful. I will assume that, besides other additional requirements (e.g. the stipulation of causal mechanisms), a *bona fide* explanation must always contain a logical, deductive, core. The no-miracle argument then relies on the following valid inference:

Argument I

1. If theory T is (at least approximately) true, T correctly predicts the phenomena (in its domain).
2. The theory T is (approximately) true.

Conclusion: T correctly predicts the phenomena.

This argument is the logical part of a 'natural explanation' of theory T's predictive success if we assume the truth[2] — or at least the approximate truth — of T (premise 2).

Besides predictive success, there is a second, sometimes neglected, notion of success involved in Putnam's argument. The (approximate) truth of predictively successful theories explains 'the way scientific theories succeed each other'. What is at stake here is not the predictive success of a particular theory, but the observationally ascertainable fact that theories are succeeded by better theories which are predictively more accurate and of broader scope. I will call this kind of success *progressive*. A 'natural' explanation of this progressive success is that historical scientific theories form a series of approximately true theories in which each member captures a larger part of the truth than the previous one. The deductive part of this explanation runs:

Argument II

1. If theory T_{i+1} is more truc than theory T_i, then theory T_{i+1} is (probably) predictively more successful than theory T_i.
2. Theory T_{i+1} is more true than theory T_i.

Conclusion: theory T_{i+1} is (probably) predictively more successful than theory T_i.

Putnam proposes a third, *methodological*, argument for SR: 'The laws of a previous theory are derivable "in the limit" from the laws of the latter theory'. This counts as a methodological rule which scientists, Putnam claims, actually follow. Let's call this rule, 'Rule L', which can be formulated thus:

Rule L : Construct a new theory T_{i+1} in such a way that it permits the derivation, in the limit, of the laws of the former predictively successful theory T_i (in the relevant empirical domain).

Scientific realism is offered as a *justification* of rule L and, by the same token, as an *explanation* of the actual (and rational) behaviour of scientists.

If successfully predictive theories are at least partially true, there is no reason to ever totally abandon them. On the contrary, new theories must be devised in such a way that they incorporate, by means of a limiting process, the laws of the older approximately true theories. Scientific realism provides a *rationale* for this method of constructing new theories as the most promising strategy to achieve continuously increasing predictive success and, at the same time, it offers an explanation of the behaviour of scientists who, in their observed practice, consciously comply — at least more often than not — with rule L.

Argument III

1. If theory T_i is (approximately) true, a new theory T_{i+1} must be constructed in accordance with rule L.
2. The predictively successful theory T_i is (approximately) true.

Conclusion: scientists must construct a new theory T_{i+1} in accordance with rule L.

This explanation of the actual behaviour of scientists is presented as an explanatory success of scientific realism itself which endows rule L with a 'methodological significance' it would otherwise lack. The rejection of scientific realism would render the alleged success of rule L a stroke of sheer luck or the result of divine intervention. Yet, it is not claimed here that shunning scientific realism would make the actual behaviour of scientists miraculous. Simply, their practice would be devoid of justification or would be rendered, perhaps, irrational or even masochistic since obeying rule L is 'often the *hardest* way to get a theory which keeps the old observational predictions' (1978, p. 20). Notice also that compliance with rule L doesn't fully guarantee but only increases the chances that T_{i+1} will be successful beyond the empirical domain of T_i and also that T_{i+1} will predict novel facts. As Musgrave (1988, p. 235) points out, no particular emphasis is laid by Putnam on a theory's ability to predict new, hitherto unexpected, facts. But Putnam at least acknowledges that this strategy leads to novel facts since he approvingly quotes Boyd and, taking the example of the conservation of energy, he says: 'That scientists try to do this [following Rule L] (…) and that this strategy has led to important *discoveries*[3] (…) is also a fact' (1978, p. 20). In other words, theoretical conservatism, which finds its *raison d'être* in scientific realism, pays.

Finally, Putnam states that scientific realism as an explanation of the predictive success of science has the epistemic status of an 'over-arching scientific hypothesis'. SR can be defended as a naturalistic position if it counts as a genuinely scientific explanation of the predictive success of scientific theories. Putnam concludes 'But then it is left obscure what realism has to do with theory of *truth*'. The nature of truth (and reference) certainly is on top of Putnam's philosophical agenda and the main reason why he is interested in the 'no-miracle argument' in the first place. His second Locke lecture is devoted to 'bring out what the connection is between explaining the success of knowledge and the theory of truth' (1978, p. 19). My claim is that no scientific connection exists between success and truth and that, consequently, the no-miracle argument does not comfort SR if SR is regarded as a scientific hypothesis as the naturalist recommends.

3. TRUTH DOES NOT SCIENTIFICALLY EXPLAIN

Michael Levin (1984) has argued that truth is not a mechanism and is devoid of any explanatory role with respect to the success of a theory. I agree with him to the extent that no scientific explanatory role can be conferred on the truth of a theory. *Contra* Levin, I'll grant some, albeit weak, explanatory power to realism as a philosophical, but not scientific, hypothesis.

Any acceptable explanation of phenomena by means of a scientific theory will involve a logically valid deduction of a proposition describing the phenomena from some premises belonging to the theory (and possible additional premises) and will also have to comply with additional requirements.[4] (For example, one may ask that some premises describe a causal process which brings about the observed effects). The details of a full-fledged account of explanation need not concern us here. It is sufficient for our purpose to assume that a scientific explanation relies on a logical inference plus additional (causal, say) requirements (R) which are not purely logical or formal conditions. Suppose that a theory T provides an explanation of some phenomena. This implies that it is possible to deduce from T propositions $e_1, e_2, ..., e_n$ which describe those phenomena. We can then construct the following argument:

Argument A

1. T logically implies $e_1, e_2, ..., e_n$

2. T

Conclusion: $e_1, e_2, ..., e_n$

Argument A is just a reformulation of Argument I above. By Tarski's (1944) criterion of truth ('Snow is white' is true (in language L) if and only if snow is white), Argument A is equivalent to:

Argument B

1. If 'T' is true, the empirical logical consequences '$e_1, e_2, ..., e_n$' of T are true.
2. 'T' is true.

Conclusion: The empirical logical consequences '$e_1, e_2, ..., e_n$' of T are true.

Now (as always) it is important to carefully distinguish between the levels of language (level 1) and metalanguage (level 2):

Level 1: $T \Rightarrow e_1, e_2, ..., e_n$

(Meta) Level 2: 'T' is true $\rightarrow$ '$e_1, e_2, ..., e_n$' are true

Suppose a genuine scientific (causal, say) explanation of the phenomena is provided by the theory T (level 1), as indicated by the double arrow '$\Rightarrow$' which is stronger than the bare logical (strict) implication '$\rightarrow$'. By semantic ascent, we certainly get a logical implication '$\rightarrow$' at the meta-level (level 2). But this is not sufficient to confer any explanatory respectability on truth, irrespective of one's preferred account of explanation.[5] Semantic ascent doesn't permit transference of the explanatory virtues of theory T with respect to phenomena at level 1 to the statement 'T is true' with respect to the truth of the empirical consequences at level 2. Tarski's criterion of truth is unable to carry the double arrow to the meta-level and to endow truth with explanatory power. Granted, we get a logical explanation of the truth of the conclusions if they logically follow from true premises. But this doesn't suffice to obtain a scientific explanation of the truth of the conclusions since the additional requirements R are not necessarily present at level 2. In other words, second-order (level 2) explanations don't belong to the same kind as first-order (level 1) scientific explanations.

We have thus shown that the semantic property of truth that a theory T possesses in virtue of reality being the way that T asserts it to be lacks scientific explanatory power. According to the realist a theory T is true iff there is some relation of correspondence *C* between the theory and the reality R, i.e. if T*C*R. A theory is successful when the same kind of relation *C'* obtains between the empirical consequence e and the phenomenon p, i.e. if and only if e*C'*p. Semantic ascent shows that if T*C*R obtains, so does e*C'*p; but it doesn't show that T*C*R explains e*C'*p. It may be objected here that what explains T's success is not its semantic property of being true, but the fact that reality is the way that T asserts it to be.[6] And if we construe, as the realist does, truth as some sort of correspondence between reality and theory, the fact that reality is the way that T asserts it to be is just the fact that the relation *C* holds and that the theory has the semantic property of being true.

However, even if we admit that there is a specific objective external reality, i.e. something which exists independently of us, and of which T is true in a correspondence sense, this doesn't scientifically explain T's success. Why? Simply because the actual correspondence between T and what it represents is not a scientific (functional, causal etc.) factor. What kind of functional and *a fortiori* quantifiable connection could be ascertained between truth and success, i.e. between correspondence at the theoretical level and correspondence at the empirical level? As far as I can see, none. On the other hand, we may concede that the correspondence between a theory and some reality provides some explanation of the truth of its empirical consequences, i.e. its success. What sort of explanation is this? It may perhaps count as a philosophical (meta) explanation of the theory's success, an explanation which works only if truth is taken as correspondence. Although an investigation of the nature and force of this explanation would surely be worth pursuing, it will not be attempted here. Let me just say that it presupposes a correspondence theory of truth; it can't be used to substantiate it. After all the success of science could be left without explanation and we can't just postulate correspondence truth to treat us with an explanation: some independent and philosophical argument in favour of correspondence truth is needed. At any rate, such a philosophical explanation doesn't add anything to the credentials of a theory on top of the grounds which can be adduced in favour of the specific claims (the existence of electrons, say) made by a theory. Paradoxically, we can accept that theory T's predictive success counts as evidence in favour of its (correspondence) truth (a view which I defend) and at the same time maintain that the truth of T is otiose with respect to any purported scientific explanation of its success.

We may add that the lack of relevance for truth in scientific explanations is reflected by the actual behaviour of scientists. If you ask a scientist why we observe such and such lines on a photographic plate, he may answer that it is because they are trajectories of electrons and electrons possess the appropriate properties (charge, mass, etc.) to account for the characteristics of the lines. A physicist certainly won't reply that it is so because the theory we have about electrons is true. Offering such an 'explanation' would be taken — at best — as a joke. But this kind of rebuttal may well leave a philosopher cold, and rightly so. After all, philosophy of science disputes are not to be democratically settled by polls among scientists but by means of philosophical argumentation.

The main source of the philosopher's uneasiness with a scientific role for truth lies in the peculiar nature of the predicate 'is true'. Truth (like falsehood) applies to statements and is thus a semantic property of some of our language components. That electrons do possess the property of having a negative charge may well cause certain facts, but nothing scientifically explanatory is added by saying that it is true that electrons have that property. As Michael Levin graphically puts it:

'Why do airplanes stay up? Surely the reason airplanes stay up is not

(1) 'The pressure on the underside of a moving airfoil is greater than the pressure on its overside' is true, but rather

(2) The pressure on the underside of a moving airfoil is greater than the pressure on its overside.' (Levin 1984, p. 126).

The realist's assertion that T is true doesn't scientifically explain anything on top of what is already explained by T, namely that the phenomena described by e_1, e_2, ..., e_n occur. If so, we may as well 'let the theory speak for itself' (Levin 1984, p. 128).

Musgrave (1988, p. 247) challenges Levin by pointing out that a theory T cannot explain its own success, which is a (meta) fact about the theory T itself. Fair enough. If T explains the phenomena described by e_1, e_2, ..., e_n, T does not explain why e_1, e_2, ..., e_n are true. Mixing up the first-order and the second-order levels is bad tactics. But the question is, does the truth of T explain (scientifically!) the success of T? We saw that it does not. Musgrave further disputes Levin's contention that ' "By being true" never satisfactorily answers the question, Why did such and such belief lead to correct expectations? The answer lies elsewhere.' (Levin 1984, p. 126) Notice that now the discussion has shifted from truth to belief in truth and these are not quite the same thing. Having true beliefs can lead to successful predictions. So we can agree with Musgrave that the fact that Hopalong believed truly that there was gold in them-thar-hills plays a role in the explaining why he actually found gold there. But this does not imply that the truth of the proposition 'There is gold in them-thar-hills' is the cause — or any scientific explanatory factor you wish to resort to — of Hopalong actually finding gold in those hills.[7] Even if Levin's formulation is slightly misleading, he is right if truth is conceived as a scientific explanatory factor, even if beliefs can of course play their part in some (even scientific) explanations.[8]

Claiming that truth doesn't scientifically explain is not to say that actual genuine explanations do not have to rely on true premises, but rather that truth is not an explanatory factor of any kind (mechanistic, causal etc.). *Mutatis mutandis* we can apply to explanations the standard distinction between valid and sound arguments. A sound argument is a valid argument the premises of which are true. Likewise, an explanation will be called satisfactory if it satisfies the requirement of logical validity, plus the additional conditions (R) imposed by one's favourite account of explanation. For example, a defender of the deductive-nomological (DN) account of explanation will demand logical validity and relevance but not the stipulation of causal processes. To be correct, in addition to being satisfactory an explanation must be true; that is, its premises have to be true and the possible causal processes it postulates must be actually operating, etc. Thus, I do not deny that truth is indeed relevant to explanation. Again, what I oppose is the purported role of truth as a scientific explanatory factor for the success of a theory T.

The only interesting question for the realist is then the following: does T provide a correct explanation of $e_1, e_2, ..., e_n$? In order to give an affirmative answer to that question, the realist must show first that the explanation provided by T is satisfactory (which can be ascertained *a priori* by means of an internal analysis of T) and second that the theoretical premises in T are true. And here lies the main difficulty, and the crucial challenge for the realist: what grounds do we have in favour of the truth of T given that it gives a satisfactory explanation of $e_1, e_2, ..., e_n$? Surely, the realist can't invoke that T gives a correct explanation of $e_1, e_2, ..., e_n$ since it is precisely the truth of T which is at stake.

4. FALLACY

The architecture of the no-miracle argumentation is captured by a sequence of three arguments which all have as a second premise a proposition asserting the truth of some theory or theories. But this is exactly what the scientific realist is after: he wants to argue in favour of the truth of theories. A correct argument for scientific realism then contains the proposition, 'Theory T is (approximately) true', not as a premise but as a conclusion. In fact, in the context of a naturalistic perspective, the scientific realist argues in the following manner. First, accept that the conclusions of Arguments I, II and III are established to be true on the basis of observations. Accept also that the first premises of an 'if..., then...' form are also true; this can be shown by examining the internal structure of a theory. It surely doesn't follow that the second premises are true. Claiming so would be tantamount to committing the 'fallacy of affirming the consequent'. That scientific realists are guilty of this mortal logical sin has been a recurrent charge by antirealists (see for example Laudan 1981, 1996). An escape from this fallacy would be to contend that the no-miracle argument is an instance of a more general form of argument, namely 'inference to the best explanation' (IBE) (see for example Musgrave 1988, Psillos 1999). Indeed, Putnam claims that scientific realism is a 'natural explanation' of the success of science, and certainly better than the rival explanation proposed by idealism (or positivism or operationalism).

Actually, Putnam, Smart (1963) and others confront us with a choice: either the theory T is true or the occurrence of $e_1, e_2, ..., e_n$ is miraculous. In other words: given that T satisfactorily explains $e_1, e_2, ..., e_n$ either T is true or some miraculous state of the world obtains. Here we are faced with a competition between two explanations: one resorting to truth, and the other involving miracles, i.e. divine interventions, or Smart's cosmic coincidences. Miracles etc. are absurd. Consequently, T must be (at least partially) true.

Before we examine how well the NMA fares at the (meta) level 2, let's have a look at what goes on at level 1. Typically, in science, we would contrast an explanation based on electrons (say), not to an explanation resorting to miracles (which you would not even dream to consider in scientific contexts) but to other explanations which are deemed acceptable given the scientific background knowledge of the day. An argumentation in favour of electron theory against rival theories is the physicist's business and is not a part of the no-miracle argument. Suppose that a scientific argumentation which shows the superiority (according to criteria whose nature need not bother us here) of electron theory over rival scientific theories has been proposed. Does this imply that electron theory is, at least approximately, true? Well, maybe 'yes', maybe 'no': it all depends on what force you think you are allowed to give to inference to the best explanation. If you think that the fact that electron theory is the best explanation of the phenomena implies the truth of electron theory, the answer is 'yes'. If, on the other hand, you have qualms (like van Fraassen 1980, pp. 19-22) about the truth-reaching power of IBE, the answer is 'no'. Whichever party you side with in this debate is irrelevant to the putative recourse to miracles since a 'miracle theory' is not a credible scientific explanation in the first place and can't *a fortiori* qualify as a worse (or better) explanation. Surely, contrasting the electron theory to the 'miracle theory' is not going to add any new credentials to electron theory.

Then, from a naturalistic point of view, what kind of scientific competition can possibly arise between an explanation invoking miracles (or cosmic coincidences) and an explanation resorting to truth at (meta) level 2? Evidently none, simply because neither alleged explanation is scientific. Granted, to resort to miracles is unacceptable anyway and we are left with the only alternative, namely truth. This way of reasoning is not fallacious but its strength relies on what happens at level 1. It is only to the extent that we may have good grounds to rule out competing scientific explanations except one that we can argue at the meta-level in favour of the truth of its premises.

5. THE METHODOLOGICAL SUCCESS OF SCIENCE

If what has been said above is correct, no naturalistic defence of scientific realism in general (i.e. the thesis that successful scientific theories are at least partially true) can be based on the dubious claim that truth scientifically explains the predictive and progressive success of science (taking these for granted). However, the situation may not be totally desperate for an ultimate argument enthusiast. Couldn't he capitalize on the methodological success, or effectiveness, of rule L in science? This line of reasoning developed by, among others, Richard Boyd is also pursued by Putnam; '(...) scientists act as they do because they believe (1) that terms in a mature science typically refer and (2) the laws of a theory in a mature science are typically approximately true and their strategy works because (1) and (2) are true' (1978, p. 21). At this point, Putnam claims, the 'causal-explanatory role' of the notions of 'truth' and 'reference' enters the field of epistemology. Scientific realism is offered as an 'explanation of the behaviour of scientists and the success of science' (1978, p. 21).

Before examining the issue of the alleged effectiveness of rule L, i.e. its ability to lead to the construction of predictively successful new theories, let us first address the problem of explaining the behaviour of scientists. (In what follows, it will be crucial to bear in mind a clear-cut distinction between (a) the truth of T and (b) the belief in the truth of T.) Assume that scientists typically believe that (mature) theories are approximately true. Given their belief, if scientists aim for true theories as the realist alleges, it certainly is rational for them to try to incorporate 'in the limit' previous theories into new, more encompassing theories. In brief, belief in the truth of some theories explains why scientists do use rule L:

Argument LB (methodological argument from belief)

1. Scientists believe in T_i.
2. Belief in T_i gives a good reason to use rule L.

Therefore: scientists use rule L.

Does the realist's assumption of the (approximate) truth of mature theories imply the truth of premise 1 in argument LB and thus explain (and legitimate as rational) the behaviour of scientists? Certainly not, simply because the link between the truth of T_i and the belief in the truth of T_i is found wanting. To make this reasoning go through, one has to assume that the event or fact (whose factual nature is peculiar) that T_i is true is a cause and that its effect is another event or fact which consists in the belief, by scientists, in the truth of T_i. This certainly is a naturalistic contention; but it is highly implausible because it clearly is not enough (nor is it necessary) that T_i be true to generate the belief in its truth. No doubt, the formation of beliefs is a highly complex matter, which involves the interplay of many factors. As of today, no general scientific theory of belief formation is in (even remote) sight. Therefore, one has to argue in favour of at least the possibility of a causal role for truth in belief formation. Such a causal role for truth seems problematic not only because of the peculiar nature of the predicate 'is true ', but also because believing in proposition *p* entails believing that *p* is true: truth seems to be part of the problem of belief formation, not of its solution. Close examination of these points would lead us too far astray. Nevertheless, we can safely conclude that the burden of a convincing case in favour of the causal role of truth in the context of this formulation of the no-miracle argument rests on the shoulders of its advocates. As things stand, SR can't be credited with an explanation of why scientists believe in their theories and consequently use rule L. Concern with fallacies aside, no plausible case for SR can be derived from its ability to account for the behaviour of scientists since SR is devoid of any explanatory force in that respect.

Belief in *p* doesn't imply that *p* is true. Surely, Putnam couldn't fail to have realised that, could he? Nevertheless, it is puzzling to read:

> If I believe principles (1) and (2), then I know that the laws of T_i are (probably) approximately true. So T_{i+1} must have a certain property — the property that the laws of T_i are 'approximately true' when we judge from the standpoint of T_{i+1} — or T_{i+1} will (probably) have no chance of being true. Since I want theories that are not just 'approximately true', but theories that have a chance of being true, I will only consider theories, as candidates for being T_{i+1}, which have this property — theories which contain the laws of T_i as a limiting case (...) In fine, my knowledge of the truth of (1) and (2) enables me to restrict the class of candidate-theories I have to consider, and thereby increase my chance of success. (Putnam 1978, p. 21)

At this point, I must confess that I fail to grasp Putnam's inference that my (or a scientist's) belief that the laws of a theory are (approximately) true, implies that I know that the laws of the theory are true. Whether we adopt the classical equation 'knowledge = justified true belief' or not, it is plain that belief doesn't imply knowledge or truth. At any rate, mere belief in T_i

provides the scientist with sufficient justification to restrict his investigation to possible new theories T_{i+1} which have the property of containing T_i in the limit. Assuming that scientists are coherent in their beliefs, we do have an explanation of why scientists use rule L.

But this can't be Putnam's argument, of course. Even if belief in the truth of mature scientific theories makes the use of rule L rational, there is no guarantee that applying rule L will generate success. Let's then move on to the issue of the effectiveness of rule L. A quick explanation of the scientists' practice would be that they use rule L because it is, in fact, effective: it has so far proven fruitful and has led to the construction of predictively successful new theories. Fine; but realism is not needed for such a justification of the behaviour of scientists. Common sense wisdom suffices: use rules that work! A plausible rejoinder to this would be that SR explains why rule L is effective and, consequently, why scientists abide by this rule. The argument runs as follows.

Argument LT (methodological argument from truth)

1. T_i is (approximately) true.
2. If T_i is (approximately) true, the use of rule L will (more probably) lead to a theory T_{i+1} which is also true (and thus predictively successful).

Therefore: rule L is, more often than not, effective (i.e. leads to predictively successful theories).

Accept for a moment that T_i is (approximately) true. Does this imply that a theory T_{i+1}, of which T_i is a limiting case, has more chances of being true than a theory, which doesn't satisfy this constraint? Yes, indeed, since T_{i+1} will at least be partially (approximately) true in the empirical domain of T_i. But what the realist needs is an argument in favour of the truths of T_{i+1} and T_i. The effectiveness of rule L hinges, by definition, on the empirical success of T_{i+1}: rule L works only if T_{i+1} yields correct predictions. Consequently, to infer the truth of T_{i+1} from the effectiveness of rule L is tantamount to inferring truth from predictive success. Since the no-miracle argument relies on the assumption that truth scientifically explains, and since this assumption is ungrounded, no convincing case in favour of SR can be derived from its alleged capacity to account for the empirical success of a given theory. The fact that rule L has contributed, sometimes, to 'important discoveries' is thus powerless to give support to SR: the naturalistic realist is sent back to the starting-blocks.

6. AGAINST NATURALISED SCIENTIFIC REALISM

Naturalists like Putnam, Boyd (1990), Psillos, among others, take the no-miracle argument to be the core ingredient of a scientific realist position interpreted as an 'over-arching' empirical hypothesis. This is hardly surprising. If scientific realism is to be construed as a scientific statement ('our best theories are at least approximately true') on a par with other scientific hypotheses, it must be argued that scientific realism does explain some facts. The no-miracle argument is precisely designed to show that scientific realism provides a scientific explanation — and the only one — of the fact that science is, by and large, successful.

First, notice that the nature of this fact is peculiar. The facts that science deals with are rather well-defined occurrences which are delimited in space and time: e.g., this light ray is bent when passing from air to glass; there are geological similarities between the East coast of South America and the West coast of Africa; the Galapagos finches have different beak shapes; far away galaxies recede from each other, etc. Here, the alleged broad success of science is a 'fact' whose precise characterisation is elusive. What counts as a mature theory? How can we accurately measure the success of a given theory? How long a period in history shall we consider? It is true that the precise characterisation and assessment of scientific facts are sometimes problematic. But here it seems that the characterisation of the 'fact' of success will not depend on scientific considerations only but will require an in-depth conceptual — and philosophical — analysis of the relevant notions. Whatever the difficulties we may face attempting to precisely characterise the 'fact', it is at least clear that such a characterisation will depend on how we evaluate the success of an individual theory.

Second, the 'fact' of success does not concern first level facts but theories: it is a second level fact or meta-fact. This idiosyncratic nature of the 'fact', even if we grant its empirical nature, complicates the search for a scientific (e.g. causal) explanation of it. Even if we concede that a particular theory T correctly explains some phenomena and is true, this is not sufficient to ground the conclusion that a scientific explanation of its success is readily available. Semantic ascent won't do the trick since, as we saw, it only preserves a logical, and not necessarily explanatory, connection between success and truth.

As of today, therefore, no scientific explanation of the success of science is available. And it is hard to imagine what form such an explanation might take. This need not bother us since this task weighs heavily on the naturalist's shoulders, but not on ours. If the success of science is to be explained at all,[9] the failure of the no-miracle argument demonstrates that a scientific explanation of success would have no bearing on the plausibility of scientific realism. Scientific realism is, as Putnam's presentation of the no-miracle argument clearly shows, an epistemological thesis which has to do with the trust we are entitled to put in science and which concerns the reach of scientific theories. For a scientific realist, it is rational to believe that our well-confirmed theories are (approximately) true in a correspondence sense, which is that they (approximately) correspond to an external, mind-independent reality. In other words, our best scientific theories provide knowledge on how parts of reality actually are and behave. Putnam's no-miracle argument was exactly designed to provide a scientific reason to believe in the truth of our successful theories. Is the failure of this project fatal to SR? I think not. But the scientific realist must jettison naturalism and give epistemology its due. To philosophy he must turn!

What could qualify as good reasons to believe that a given theory is nearly true and that its theoretical parts correspond to independently existing, though perhaps unobservable, entities? This is the philosophical challenge that confronts us. Though such an enterprise obviously goes beyond the scope of this paper, let me briefly delineate how it might be carried out.[10] The most promising way, it seems to me, is to start from common everyday experience, as classical empiricists recommend, and to analyse the grounds for our beliefs in the existence of ordinary objects (cats, mice, cheese and so on) and statements about those objects. Granted, in doing so, we attribute to everyday experience some normative force. But unless we are ready to embrace extreme scepticism, what else can we do? Surely, the broader the variety and scope of available evidence, the better support we have for our beliefs. Novel predictions are important in that respect, but they simply contribute — quantitatively and not qualitatively — to provide more grounds for belief without providing a safe haven against error. The next — decisive, but intricate — step will be to determine whether the same kind of rational grounds can be used to legitimate our theoretical beliefs in the existence of some unobservable entities postulated by some scientific theories and the truth of some scientific laws.[11] Such a vindication of SR would be a genuinely philosophical, and not a scientific, achievement.[12]

Université Catholique de Louvain

REFERENCES

Boyd, R.: 1990, 'Realism, Approximate Truth and Philosophical Method' in C. W. Savage (ed.), *Scientific Theories, Minnesota Studies in the Philosophy of Science* Vol. 14, University of Minnesota Press, Minneapolis, 1990.

Fine, A.: 1986, 'Unnatural Attitudes: Realist and Instrumentalist Attachments to Science', *Mind* 95, 149-179.

Ghins, M.: 1992, 'Scientific Realism and Invariance', *Philosophical Issues Vol. 2*: Rationality in Epistemology, Ridgeview, California, pp. 249-62.

Ghins, M.: 2000, 'Empirical versus Theoretical Invariance and Truth' followed by a commentary by B. van Fraassen, *Foundations of Physics* 30, 1643-1655.

Giere, R.: 1988, *Explaining Science: A Cognitive Approach*, University of Chicago Press, Chicago.

Laudan, L.: 1981, 'A Confutation of Convergent Realism', *Philosophy of Science* 48, 19-49.

Laudan, L.: 1996, *Beyond Positivism and Relativism*, Westview, Boulder.

Leplin, J.: 1997, *A Novel Defence of Scientific Realism*, Oxford University Press, Oxford.

Levin, M.: 1984, 'What Kind of Explanation is Truth? ' in J. Leplin, *Scientific Realism*, University of California Press, Berkeley, 1984, 124-139.

Musgrave, A.: 1988, 'The Ultimate Argument for Scientific Realism ' in R. Nola, *Relativism and Realism in Science*, Kluwer, Dordrecht, pp. 229-252.

Psillos, S.: 1999, Scientific Realism : *How Science tracks Truth*, Routledge, London.

Putnam, H.: 1975, 'What is Mathematical Truth?' in H. Putnam, Mathematics, Matter and Method. *Philosophical Papers Volume I*, Cambridge University Press, Cambridge, pp. 60-78.

Putnam, H.: 1978, *Meaning and the Moral Sciences*, Routledge, London.

Sankey, H.:2000, 'Methodological Pluralism, Normative Naturalism and the Realist Aim of Science' in R. Nola and H. Sankey (eds.), After *Popper, Kuhn and Feyerabend*, Kluwer, Dordrecht, pp. 211-229.

Smart, J. J. C.: 1963, *Philosophy and Scientific Realism*. Routledge, London.

Tarski, A.: 1944, 'The Semantic Conception of Truth and the Foundations of Semantics', *Philosophy and Phenomenological Research* 4, 341-376.

Van Fraassen, B.C.: 1980, *The Scientific Image*, Clarendon Press, Oxford.

Van Fraassen, B.C.: 1989, *Laws and Symmetry*, Clarendon Press, Oxford.

NOTES

[1]Putnam develops his argumentation within the framework of the syntactic view of theories. Like others (van Fraassen 1980, 1989; Giere 1988) I favour the semantic, model-theoretic approach. But I don't think the issues discussed here are sensitive to this debate. Acceptable scientific models must lead to correct empirical predictions. And any account of explanation must be compatible with the deduction of predictive sentences from the "laws" which truly describe the model, plus boundary conditions, etc.

[2]As is well-known, Putnam is more concerned with reference (or satisfaction), of which truth is a particular case. But we will limit ourselves to truth since the no-miracle argument can be used in almost the same form to defend the existence of the reference of theoretical terms.

[3]My italics. Discoveries are novel facts.

[4]There exist explanations in which the conclusion doesn't follow logically, but only probabilistically, from the premises (Hempel 1962). But the point I want to make applies to

those as well: the truth of the premises doesn't scientifically explain the degree of probability of the conclusion. Moreover, phenomena that are made only probable by the conditions described in the premises seem to appear less 'miraculous' if those conditions do not obtain than phenomena which follow from these conditions.

[5]Since the additional requirements for having an explanation are not merely logical, this holds even if we allow for some stronger formal implications.

[6]I'm especially grateful to Howard Sankey on this issue. See also the end of Sankey (2000).

[7]Notice also that the question 'Why did Hopalong find gold in them-thar-hills?' presupposes that there actually is gold there.

[8]For example, fear and shivering can be caused by some (false or true) beliefs.

[9]I'm sceptical of instrumentalist (Fine 1986, 153-154) and Darwinian (van Fraassen 1980, pp. 39-40) explanations. Fine's explanation is circular (see Psillos 1999, pp. 92-93) and the Darwinian account is unable to explain why a *particular* theory T is successful, as has been often pointed out.

[10]For more on this see Ghins (1992, 2000).

[11]Just as ordinary existing objects satisfy formal conditions of invariance (constancy) and material conditions of actual presence in observation, we can demand that only theoretical objects which obey formal symmetry requirements and are connected with actual measurements can be reasonably thought to be existing.

[12]I wish to thank Otavio Bueno, Jacob Busch, Silvio Chibeni, Brian Ellis, Tim Lyons, Robert Nola, Howard Sankey and Bas van Fraassen for stimulating discussions or comments on the topic of this paper.

JOHN WRIGHT

SOME SURPRISING PHENOMENA AND SOME UNSATISFACTORY EXPLANATIONS OF THEM

1. INTRODUCTORY REMARKS

A feature of science that makes it of philosophical interest is its ability to *do* certain things. One thing that science seems to be able to do is predict the natural world, more specifically, to successfully predict *novel* phenomena. Another ability that science *perhaps* has is that of being able to give us true descriptions of relatively *inaccessible* parts of the natural world, such as the interior of the atom and distant parts of space and time; although it is of course controversial whether it really has that ability. Yet another ability that — it will here be argued — science has is to lead us, more or less *a priori*, to theories that subsequently turn out to be empirically successful.

The problem that forms the primary question addressed in this paper is, *if* science does have these abilities, how is its possession of these abilities to be explained? It is useful for later reference if we number the phenomena with which we will be concerned:

Phenomenon One: The ability of science to produce theories that have novel predictive success.

Phenomenon Two: The ability of science to give us true descriptions of parts of the world that were unobservable or inaccessible at the time that theory was first advanced. These unobservable or inaccessible parts of the world may include distant parts of space and time, entities that are too small to see, or entities that were, for some other reason, unobservable when first postulated.

Phenomenon Three: The ability of science to lead us, on more or less *a priori grounds*, to some theories that subsequently turn out to be empirically successful.

S. Clarke and T.D. Lyons (eds.), Recent Themes in the Philosophy of Science, 139–155.

2. THE GENUINENESS OF THE PHENOMENA

A theory enjoys *novel* predictive success if it successfully predicts phenomena different from those on the basis of which the theory was initially formulated. The following are well-known examples of novel predictive success: the prediction that if enough Uranium-235 is brought into very close proximity, an explosion 'brighter than a thousand suns' will result; the prediction, made by the eighteenth century mathematician Poisson, that if a round object is placed in a beam of light, a small white spot would be observed in the centre of the shadow; and the prediction made by the Special Theory of Relativity that if two extremely accurate clocks are first synchronised, and one flown around the world while the other remains stationary, the former would be found to be slightly slower than the latter. It will here be assumed that it is indeed a genuine fact about science that it sometimes succeeds in predicting novel phenomena.[1]

It is important to note here that the thesis that scientists have managed to select *some* true theories which make claims about inaccessible or unobservable parts of nature is a weaker claim than the thesis of Scientific Realism, at least as that thesis is normally understood. I will take Scientific Realism to be thesis that mature scientific theories are typically approximately true, or — what is certainly not the same thing — that the terms of mature scientific theories typically refer.[2] Thus construed, Scientific Realism is a claim about *typical* mature scientific theories, and therefore presumably implies that at least *most* mature scientific theories are approximately true, or referential, or both. But if achieving truth is sufficiently improbable or surprising, even a *handful* of cases of scientists postulating true theories about inaccessible or unobservable would require explanation. We can also note here that there surely are some cases of scientists postulating entities which were inaccessible at the time, but which subsequently turned out to exist and to possess the properties that had been attributed to them. One example of this is the postulation, by Adams and Leverrier, of the planet Neptune.[3] At the time that Adams and Leverrier made their calculations, the planet Neptune had not been observed. The prediction that such a planet exists was therefore a novel prediction of the Newtonian model of the solar system. But, given that spacecraft have now flown past and photographed it, Neptune's existence is surely now beyond any serious dispute.

The controversial 'zone' of debate between realists and anti-realists with respect to *ontological* doubt is constantly changing. Here are some things that were once the subject of ontological doubt, but which are no longer — or, at least, they are less dubious than they once were: the (approximate) sphericity of the Earth, the fact that the Earth orbits the Sun, the rings of Saturn, the planets Neptune and Pluto, the status of the fixed stars as entities like our Sun, galaxies, organic cells, bacteria, viruses, molecules, atoms, protons, electrons, neutrons. Many theories that were once subject to ontological doubt are now much less so. But this is a phenomenon that requires explanation: how is it that theories which, when initially advanced, postulated entities that were not observable, subsequently turned out to be right about the existence of those entities? This is one type of Phenomenon Two.

In summary, it seems fairly uncontroversial that there are at least some cases of Phenomenon Two, eg., the postulation of Neptune and the germ theory of disease. But if, with the scientific realist, we wish to explain novel success by saying that successful theories are (at least approximately) true, then there will be many more cases.

The third phenomenon with which we will be concerned is the fact that some theories which we find preferable on *a priori* grounds subsequently turn out to be highly empirically successful. Conservation laws are a good example of this, as are Newton's three laws of motion and his law of gravitation. In this section it will be argued that, although there is a sense in which these theories are accepted *a priori*, they have turned out to be surprisingly empirically successful.

First, we need to note that there are (at least) two different notions of '*a priori*'. On what I take to be the standard notion, a proposition is knowable *a priori* if and only if the only experience necessary to know that it is true is that which is necessary to know the meanings of the terms in it. I will call this '*a priori* knowledge', or '*a priori* certainty'. But there is another type of *a prioricity*, which is naturally called '*a priori* acceptance' or '*a priori* belief'. A proposition is accepted *a priori* if and only if we find ourselves accepting it even though there is no evidence for its truth in our experience. It will be argued that there are some types of scientific theories that we have an *a priori* tendency to *accept* or *believe*, but which subsequently turn out to enjoy empirical success.

It might be objected that if things always became more massive after mixings or dissolvings, or interactions of some other kind, then, in the long run, things that underwent many such interactions would eventually become noticeably heavier. But it is easy to think of other possible laws that would prevent any increase or decrease in mass ever becoming detectable. Perhaps things slightly increase their mass in the first, third and fifth interactions, while decreasing on the second, fourth and sixth and so on. Or interactions on Monday, Wednesday and Friday bring about a slight but undetectable increase in mass, while those on Tuesdays, Thursdays and Saturdays bring about a decrease. On Sundays, mass is conserved. Again, there is clearly an infinite number of ways in which mass could be slightly increasing or decreasing all the time, and all these different ways are compatible with our everyday experience.

Despite the fact that there are many laws about how matter behaves that are compatible with our everyday experience, we find the idea that it is conserved much more plausible than the alternatives, such as a law stating that matter alternately increases and decreases by an imperceptible amount. '*Why* would it increase? Where would the extra matter come from? How would it arrange itself to increase and decrease *alternately*?' These are the questions we would ask ourselves. And in the absence of an answer to these questions we would take the alternative hypotheses to be far less plausible than the idea that matter is conserved. So, I take it as uncontroversial that we would regard the hypothesis that matter is conserved as being much more likely or reasonable than the others.

But now, our conviction that it is more likely that matter is conserved is clearly not supported by anything in our *experience*. While our experience is *compatible* with its being conserved, it is also compatible with very many other hypotheses. We regard it as more likely that matter is conserved than that it, say, alternately increases and decreases by a very small amount; but this belief is not derived from, nor indeed given any support by, our experience. Our tendency to prefer the hypothesis that it is conserved is, therefore, an *a priori* preference. Note that it is not being claimed here that it is *known a priori* that matter is conserved. Rather, what is accepted *a priori* is the preferability of 'Matter is conserved' to the other generalisations that are also compatible with our everyday experience.

One type of answer that has had its adherents is that the law of the conservation of mass does not really make a claim about how reality is at all; it is, rather, merely a conventional truth.[4] But this point of view is now difficult to take seriously. According to the theory of relativity, it is false that matter is always conserved: it is sometimes converted into energy. Moreover, it has been observed that there has been an apparent decrease in mass after atomic explosions, and in nuclear reactors. When interpreting these observations, scientists have not concluded that the missing mass must still exist somewhere undetected; instead, they conclude that the mass has not been conserved at all, but has been converted into energy. That scientists interpret the results in this way is incompatible with the idea that the law of conservation of mass is a conventional truth. Claiming that it is a conventional truth will not, then, adequately explain why it continues to be confirmed when we pass from everyday observations to observations made in the chemical laboratory. The falsifiability of the law of the conservation of mass assures us that it is a synthetic statement with empirical content. But then its success, once we move from everyday experience to the chemical laboratory, becomes all the more surprising. Why should an empirical, synthetic statement, which we find *a priori* plausible, continue to receive confirmation when we test it at a more precise level, when other statements, equally confirmed by everyday experience but not *a priori* plausible, are refuted? Does this mean *a priori* plausibility actually makes subsequent empirical success more likely? But why should this be so? We are confronted here with a phenomenon that requires explanation.

3. SOME UNSATISFACTORY EXPLANATIONS OF THE PHENOMENA

(a) *Are the successes due to luck?*

Let us begin by considering how we might explain the novel predictive success of science. One natural suggestion is that perhaps the success is simply due to good luck. Certainly, science has had some impressive predictive successes, but it has also had many predictive failures. So, perhaps the novel predictive *successes* of science are, roughly, no more than would be expected by chance.

(1) Magnetic moment of electron
Value predicted by QED: $1159652359 \times 10^{-12}$
Value obtained by experiment: $1159652410 \times 10^{-12}$

(2) Value of Lamb shift for the hydrogen atom
Value predicted by QED: 1057.864
Value obtained by experiment: 1057.893

(3) Value of Muonium Hyperfine Structure
Value predicted by QED: 4463.293
Value obtained by experiment: 4463.30235

(4) Value of Positronium Hyperfine structure
Value predicted by QED: 203.3812
Value obtained by experiment: 203.3849

(5) Value of Positronium Spectrum
Value predicted by QED: 8.62514
Value obtained by experiment: 8.6284

It should be noted that this list of the empirical successes of quantum electrodynamics is not exhaustive.[5] Now, let us consider the *a priori* probability that these agreements between observation and theory have occurred by chance. Consider the first result, that for the magnetic moment of the electron. There is a very reasonable line of argument[6] that leads us to conclude that the *a priori* probability of such agreement is not greater than approximately 5×10^{-8}. By the same reasoning, the *a priori* probabilities of the other results are 3×10^{-5}, 4×10^{-6}, 2.5×10^{-5} and 6.6×10^{-4} respectively. Assuming all these results are independent of each other, the probability of all these results being obtained is just the product of the individual probabilities; that is: 990×10^{-28}, or approximately 10^{-25}.

Clearly, this probability is very low. We can get an intuitive idea of just how low this probability is by asking how long it would take to get results this improbable by a process of random guessing. The answer: if ten million theorists were each advancing one hypothesis every second, it would take longer than the History of the Universe before the probability became greater than one half that they should obtain, by chance, results that agree this closely with observation. The idea that the predictive successes of quantum electrodynamics are just due to chance must, therefore, be rejected.

One might raise the following objection. The above argument is based on the assumption that the five results are independent of each other, but perhaps this is not so. It seems to be at least *possible* that the success of one of the tests might increase the probability that the next test will also be passed. Were this the case, it could not be concluded that the probability of the above results is the extremely low figure of 10^{-25}. This objection, however, fails: either the results are independent, in which case the figure of 10^{-25} stands, or, if they are not independent, then we are driven back to square one with respect to the problem of explaining the predictive success of science. Consider the following conditional claim (C):

> *If* the value of one of the quantities (say, the magnetic moment of the electron) is X, then there is some better than chance probability that the value of another of the quantities (say, the Lamb shift of the hydrogen atom) will be Y.

What is the status of this claim? Plainly, it is not an *a priori* or analytic truth. If it is true, and known to be true prior to the empirical determination of the Lamb shift, it can only be because some *theoretical* claim is known to be true. Therefore, from a point of view that is agnostic with respect to the truth or falsity of any theoretical claim, the value obtained for one of the quantities (say, the magnetic moment of the electron) is independent of the value obtained for any of the other magnitudes. Hence, so long as we remain agnostic about the truth or falsity of any theoretical claim, the figure of 10^{-25} is a fair estimate of the probability of obtaining theoretical predictions that agree with that degree of accuracy with the experimental findings.

Another objection that may be raised against the argument given here for the extreme improbability of the predictive successes of science is that these five successes represent only a part of the picture. Although quantum electrodynamics has not yet been falsified, the history of science is full of cases of theories that led to predictions that turned out to be wrong. If the *only* predictions ever made by science were these five, then the fact that subsequent experimental tests confirmed them to this degree of accuracy would indeed be astonishing, but the fact that there have also been innumerable predictive failures makes these five successes much less surprising, and much less improbable.

(b) *Does Scientific Realism explain the predictive success of science?*

Scientific realists, of course, hold that the success of scientific theories is to be explained by saying that they are true, or, in some sense close to the truth.[7] It is, of course, a highly controversial issue in the philosophy of science whether the best explanation of the predictive success of science is the (approximate) truth of scientific theories. But, if we say that the success is due to (approximate) truth, we are confronted with a new problem: how have scientists managed to hit upon true theories? Scientific theories are in many ways about *inaccessible* parts of the world. Some theories are about entities, such as atoms and mesons, that are too small to see. Other theories are about entities, such as radio waves, which are at least as big as many entities we can see but which cannot be directly detected. Others make claims about objects, such as quasars and black holes, too far away from us to be directly detected with our senses. Still other theories can make claims about wholly familiar things, but the claims are of such a character that common sense would have no idea how to go about verifying or falsifying them, and indeed the claims can almost seem nonsensical: for example, General Relativity asserts that gravity is curved space-time. Finally, *all* theories that are explanatory, universal generalisations make claims about a potential infinity of states of affairs, about distant points of space, and about the distant past and distant future. More generally, our theories make claims about relatively *inaccessible* parts or aspects of the world. So, if, as the realist may claim, we have managed to make true claims about these less accessible parts of the world, we must ask how we have managed to do this.

Can we perhaps say that we have managed to hit upon (approximately) true theories by chance? It is easy to show, however, that the probability of doing this is comparable to the probability of hitting upon correct empirical predictions by chance. For any possible set of empirical results, there will be *some* 'theory' or set of theoretical statements capable of explaining those results; although some of those 'theories' may be very complex, *ad hoc* or highly implausible. For example, any conceivable set of empirical results can be explained by saying God had a specific intention to produce precisely those results. Any other possible set of results could be explained by attributing a different intention to God. For any possible set of results, there will be a theory attributing the corresponding *intention* to God, so the number of possible theories is at least as great as the number of possible empirical results. The chances, therefore, of hitting upon, by random guesswork, a theory that turns out to have correct empirical consequences is at least as great as the chances of hitting upon, by random guesswork, the true empirical statements themselves. But since a theory is *true* only if it has correct empirical consequences, it follows that the chances of hitting upon a true theory are at least as low as the chances of hitting upon, by random guesswork, true empirical statements. And we have seen that the chances of that are astronomically low.

One natural response to this argument is to point out that it includes, amongst the explanatory theories worthy of consideration, such scientifically unusual hypotheses as 'This result was due to an intention of God'. Perhaps, it may be objected, if we are allowed to include such unusual hypotheses amongst the set of all hypotheses worthy of consideration, then the set of all such hypotheses will be at least as large as the set of all possible empirical results, and so the chances of hitting upon a true theory will be at least as low as the chances of hitting upon a correct set of empirical results. But if we are to restrict the set of hypotheses worthy of serious consideration to some *subset* of all possible hypotheses — to, say, the set of hypotheses that are scientifically 'plausible' or simple, or non-*ad hoc* — then maybe the chances of hitting upon a true or approximately true theory are greatly increased. There are many difficulties that confront this suggestion, but here is just one: Suppose that restricting the range of admissible hypotheses to those that were simple, or plausible, or 'elegant' *did* increase the chances of our hypotheses being true. Then we are confronted with a new problem: why is it that plausibility, simplicity, etc. increase the chances of truth? And if they do, how did we manage to discover the particular kinds of simplicity, plausibility, etc. that are indicators of truth — after all, what is plausible to us might not be plausible to the Martians. These questions are considered in the next section.

(c) *The 'meta-inductive' explanation of the success of scientific methods*

In this section I will consider what is perhaps the most promising strategy for explaining how we have managed to produce theories that enjoy novel predictive success. This is the 'meta-inductive' explanation of the success of our methods.[8] It runs as follows. Scientists are, all the time, advancing hypotheses. By chance, some of these hypotheses enjoy novel predictive success. It is noticed that the ones that enjoy novel predictive success have (or tend to have) some property M. This property M might be 'simplicity', 'plausibility', 'naturalness', 'elegance', etc. Here we will just call it 'property M'. This leads scientists, when considering how to construct *new* theories, or when considering which of a number of candidate hypotheses to accept, to prefer those with property M. Consequently, future hypotheses will have (or tend to have) property M. But since property M has been shown, by past experience, to be associated with novel success, it is only to be expected that the new hypotheses that have M will also have novel success.

Now, some sort of account like this may very well be partially correct. But, as an *explanation* of subsequent predictive success it is unsatisfactory. In order to see this, we first need to get clear on precisely what must be assumed if this explanation is to work. It will be easier to do this if we explicitly spell the argument out in a number of distinct steps:

(1) Scientists have discovered, or noted, that those theories *advanced in the past* that enjoyed novel predictive success tended to have property M.

(2) Scientists hit upon the hypothesis that theories postulated in the future that have property M will tend to enjoy novel predictive success.

(3) For this reason, scientists, in their practice, have tended to favour theories with property M.

(4) And theories with property M do indeed (tend to) have novel predictive success.

---

(5) Therefore: Theories advanced by scientists (tend to) have novel predictive success.

The point that needs to be noted here is that (4) — the claim that theories with property M do indeed tend to have novel predictive success — plays a *non-redundant role* in the explanation. But if (4) is true, it is a rather surprising fact. On the proposed explanation, scientists may discover M either by a 'meta-induction' on past theories, or perhaps by just a stab in the dark. But why should property M *continue* to be a reliable guide to future predictive success when scientists come to theorise about new domains? Not that long ago, our theories were about things on the surface of the Earth — rocks, trees, animals, balls rolling down slopes, etc. And for many centuries before our theories would have been about such familiar, everyday objects. Early scientists may have hit upon some property M possessed by the theories about these familiar objects. And, on the meta-inductive account, this led them to *prefer* theories that have property M when they came to theorize about, for example, areas of space beyond the surface of the Earth, or about the very small, or about very high velocities or very high energies. On the meta-inductive account, it is *because* property M continues to be a reliable indicator of future empirical success in these new domains that our theories in these domains enjoy novel success. But if this is correct, it is rather *surprising*: it could have been that property M, which we had come to favor while theorizing about tables and chairs, trees, rocks and rivers and so on, proved to be quite unreliable as a guide to successful theories about protons, mesons, and curved space-time. And so we are confronted with the question: why should M have continued to be successful? A complete explanation of novel success — particularly in these unfamiliar domains — would need to include some kind of explanation of *why* our sense of theoretical goodness did continue to work so well.

An objection may be raised against the preceding argument. Consider a simple scientific law such as 'All silver conducts electricity'. We may have initially arrived at this law by observing samples of silver on the Earth. But we are nevertheless confident that if we were to encounter samples of silver on, say, the Moon or Mars, they would still conduct electricity. Moreover, this confidence seems entirely reasonable. So — it might be objected — it seems reasonable to assume that, if the generalization 'All theories with M (tend to) have subsequent predictive success' has received positive instances in familiar domains, it will continue to receive positive instances when we come to theorize about unfamiliar domains.[9]

However, compared to typical scientific laws, the hypothesis 'All theories with property M (tend to) have subsequent success' is very 'odd'. This can be brought out if we consider what general type of property M is likely to be. If we are confronted with a number of hypotheses, all of which can explain some body of observations, we use M to select which of them is the best. Property M is therefore likely to be something like: *simplicity* or *plausibility* or *mathematical elegance* or some similar concept. But it is a feature of these concepts that they are all closely linked to human tastes or preferences. There is ultimately no test for plausibility or mathematical elegance other than what *we find* to be plausible or elegant. Some philosophers have proposed definitions of simplicity, or algorithms for measuring it, but we appeal to our intuitions of simplicity in evaluating those definitions. Also, of course, what is used in scientific practice is scientist's own intuitive judgements concerning simplicity and so it is those intuitive judgements that have led us to success. But whether a theory subsequently receives experimental confirmation or not, or whether a theory turns out to be close to the truth or not, would seem to be something quite independent of human tastes and preferences. These things would seem to depend upon how the world 'out there' actually is. So the generalization 'All theories with M (tend to) enjoy subsequent predictive success' asserts that there is a correlation between something closely linked to human tastes and preferences and how things actually are in the universe 'out there'. This claim may be true, but if it is true it is certainly very surprising. And if it were true, we would want an explanation of this fact.

There are some other difficulties with the meta-inductive explanation of success. One is that it asks us to accept certain factual claims which are rather hard to actually believe. Consider, for example, our preference for smooth curves. It does not seem to be the case that we have acquired our preference for smooth curves by a laborious process of trial and error, that is, by first trying out all sorts of snaking and jagged curves, finding out they were not good at successfully predicting new values of the properties under consideration, and then finally hitting upon the smooth type of curves we actually do prefer. For example, we do not find school children drawing jagged or snaking curves between the points in science exercises. Smooth curves seem to be 'just common-sense'. Or again, consider our preference for simple explanations. It does not *appear* to be the case that people acquire this preference by first trying all sorts of complicated and *ad hoc* explanations, finding that such explanations are not very good at leading to successful predictions, and then developing a preference for simple explanations. On the face of it, the meta-inductive account of how we came to acquire our epistemic preferences does not seem to square with our experience.

One more shortcoming for the meta-inductive explanation. So far, we have considered whether the meta-inductive account might be able to explain how we manage to hit upon theories that subsequently turn out to have predictive success. It has been argued that it does not seem to be able to provide a satisfactory explanation of this. But it is also worth noting that it would be quite unable to explain the third of the phenomena with which we are here concerned: the empirical success of some *a priori* preferable theories. On the meta-inductive account, we would have arrived at our *preference* for conservation theories, for example, by a process of trial and error; that is, we would have tried a range of different types of theories, some of which had quantities conserved, others of which did not, found conservation theories to generally have a better record for subsequent predictive success, and so developed a preference for conservation theories over others. This is why we now prefer conservation theories. But this account has an obvious flaw. As we have already noted, all our experience is compatible with the truth of conservation laws, but it is also compatible with the truth of many other laws which say that quantities are always imperceptibly increasing or decreasing. The meta-inductive account is therefore confronted with a problem: why has this process of trial and error led to a preference for conservation theories rather than to a preference for one of the other types of theories that have been compatible with our experience?

Note that it is of no help to the advocate of the meta-inductive account to appeal to the greater *simplicity* of conservation laws. The problem that arises with conservation also arises with simplicity. Suppose it is suggested that we acquired our preference for simple theories by a process of trial and error. This would involve trying all sorts of different theories, some of which are simple, others of which are not, and finding that the simple ones tended to enjoy better rates of subsequent confirmation. The difficulty is that, just as many non-conservative laws are as compatible with our experience as conservation laws, so many *complex* theories are as compatible with our experience as simple theories. Our earlier discussion of conservation laws provides us with some examples of this. The hypothesis that matter is conserved is a lot simpler than the hypothesis that it alternately increases and decreases by 0.0000000001%. But the latter hypothesis is as compatible with our experience as the claim that it is conserved. So, on the meta-inductive account, it is at least unclear why we should have come to acquire a preference for the *simple* theory rather than some sort of non-simple theory that is also compatible with all our experience. I conclude that the meta-inductive explanation of the predictive success of science is not satisfactory.

(d) *'Evolutionary' explanations of success*

It might be suggested that some kind of 'evolutionary' explanation could be given for the success of science.[10] An evolutionary explanation might be given either at the level of individual theories or at the level of our methods. At the level of individual theories, an evolutionary explanation would go something like this: theories are formulated by a process of blind variation. Some of those theories subsequently turn out to be successful, others do not. The successful theories are (provisionally) retained in science, the unsuccessful ones are not. In this way, science eventually comes to be populated only by successful theories. However, it is plain that this account does not satisfactorily explain highly improbable novel predictive successes. If theories are formulated by a process of blind variation, how are we to account for such *a priori* improbable new empirical successes, such those of QED described in Section 3(a)?

An 'evolutionary' account might also be given of how we arrived at our tendency to prefer simple, 'elegant', conservative, etc. theories. However, this would simply be a variant of the meta-inductive explanation of success, and so subject to the same difficulties. An evolutionary account of how we acquired our tendency to prefer, say, simple theories, would naturally proceed as follows: Preferences for simple theories developed as a result of blind variation. If those preferences led us to theories that were successful, the preferences were retained, if not, they were rejected. Since simple theories have, as a matter of historical fact, led to success, our preference for simple theories is one that has been retained. However, this does not *explain* why our preference for simple theories continued to lead to success, particularly in domains different from those in which our preference for successful theories was initially acquired.

(e) *Could we explain the phenomena by hypothesizing that nature itself is simple?*

One natural way to explain science's ability to achieve novel predictive success is to claim that nature itself is simple.[11] If nature itself is simple, then simple theories would seem to stand a better chance of being true, and hence predictively successful. However, the hypothesis that nature is simple only provides us with half an explanation of the phenomena. We would also need an explanation of why we *prefer* simple theories; or, more precisely, we would also require an explanation of why the type of theories that we prefer have a property that reflects this general feature of the universe. Is it merely good luck? After all, it seems possible that the type of theories we prefer could have *not* reflected this general feature of the universe. But if we say that we *discovered* that simple theories are more likely to be true, or successful, by a process of trial and error, then we are back to the meta-inductive or evolutionary explanations, the shortcomings of which we have already investigated.

(f) *Could science itself provide us with a satisfactory explanation of its own successes?*

One might suggest that science explains the success of science. Perhaps the most obvious sciences to look to for an explanation of the ability of the human brain or mind to produce successful theories are psychology and/or cognitive science.[12] But a little reflection shows that these sciences could not provide a *complete* explanation of the phenomena with which we are here concerned. We cannot 'see directly' into, for example, the interior of the atom or the behavior of mesons or quarks. Neither can we see into the future and see directly what the outcomes of future tests of our theories are going to be. So, if we do manage to hit upon theories about, say, the interior of the atom that are true, or if we do manage to hit upon a theory that passes future tests, it is *not* because we have *directly seen* the interior of the atom, or the results of future tests, and selected the right theories on that basis. All that we have access to is data drawn from the past and present about more directly observable parts of the world. And we use criteria such as simplicity, plausibility, elegance and perhaps many other criteria to select, from amongst all the theories that are capable of explaining our data, that theory which we think is the best. If we are lucky, the theory that we decide is the best does turn out to receive subsequent confirmation: We might get subsequent confirmation that what it says about the interior of the atom is right, or it might pass novel tests we subject it to in the future. Now, plainly, a *complete* explanation of such successes would need to answer (at least) the following two questions:

> Question (1) How did the human brain or mind, when presented with some body of data, manage to formulate some theory, or find some pattern in the data, that was also simple, plausible, elegant, etc?
>
> Question (2) Why is it that the theory that was simple, elegant, plausible, etc. successfully passed novel tests to which we later subjected it? Or, why did we get subsequent confirmation that the simple, plausible, etc. theory was true or close to the truth?

Very plausibly, sciences such as Psychology or Cognitive Science are the appropriate ones to look to when answering the first question. But it is also clear that those sciences by themselves will not provide us an answer to the second question. To answer the second question we need to explain why those theories we judged to be good (those with simplicity, plausibility, etc.) turned out to subsequently be empirically successful in new ways, or to be (approximately) true. It seems any science-using explanation of this fact would need to mention, not just the theories themselves and the mental processes that gave rise to them, but also those parts of the world that the theories are about. For example, suppose we looked to science for an explanation of why the *simplest* theory of the interior of the atom subsequently passed some novel tests. In order to explain this it would not be enough to give some account of how the human brain managed to arrive at a simple theory that explained some collection of data; we would also need to explain why the predictions made by that theory concerning how the interior of the atom behaved turned out to be true. And in order to do that we would also need to appeal to our theories about the interior of the atom itself. So, sciences such as Psychology and Cognitive Science are not sufficient to answer Question (2).

4. CONCLUSIONS

In this paper we have considered three varieties of the success of science. These are (1) the ability of science to successfully lead us to novel predictive success, (2) the ability of science to give us knowledge of parts of the world that were not accessible at the time the theories about them were first advanced, and (3) the ability of some *a priori* plausible theories to receive more subsequent confirmation than their experientially equivalent rivals. It has been argued that these phenomena are genuine. We have considered a wide range of possible explanations of the phenomena, including the suggestions that these successes are no more than we could reasonably expect by chance, that Scientific Realism could explain the phenomena, that they can be given 'meta-inductive', or 'evolutionary' explanations, that they could be explained by the hypothesis that the universe itself is simple, or that sciences such as psychology or cognitive science might be able to provide the explanation. It has been argued that none of these approaches are satisfactory. We may therefore conclude that the ability of science to successfully lead to new knowledge, whether at the observational or theoretical level, in domains that go beyond those that were used to initially develop the theory, is, as yet, something that is not properly understood.

University of Newcastle

REFERENCES

Boyd, R. : 1983, 'On the Current Status of the Issue of Scientific Realism' in *Erkenntnis* 19, 45-90.

Drell, S. D.: 1979, 'Experimental Status of Quantum Electrodynamics', *Physica A: Theoretical and Statistical Physics*, 96A, pp. 3-16.

Feynman, R.: 1992, *The Character of Physical Law*, Penguin Books, London.

Friedman, F. L.: 1966, *Physics*, D.C. Heath & Co., Boston.

Laudan, L.: 1984, 'Explaining the Success of Science: Beyond Epistemic Realism and Relativism' in J.T. Cushing et al. (eds), *Science and Reality: Recent Work in the Philosophy of Science*, University of Notre Dame Press, Notre Dame, 1984, pp. 83-105.

Newton-Smith, W. H.: 1981, *The Rationality of Science*, Routledge and Kegan Paul, London.

Putnam, H.: 1984, 'What is Realism?' in J. Leplin (ed), *Scientific Realism*, University of California Press, Berkeley, 1984, pp. 140-154.

Wright, J.: 1997, *Realism and Explanatory Priority*, Kluwer Academic Publishers, Dordrecht.

NOTES

[1]For an explication of the notion of novel success in science see Wright (1997, pp. 172-182).

[2]This definition of Scientific Realism is used by Putnam. See his (1984, pp. 140-154, especially p.142). It is also close to, although something of a simplification of, the characterisation of realism given by Richard Boyd (1983, pp. 45-90, especially p. 45).

[3]It might be objected that the planet Neptune was not, even at the time of Leverrier and Adams, an *unobservable* entity, it was, rather, an *observable* entity that had not yet been observed. This may well be true, but it does not matter from our point of view. Adams and Leverrier succeeded in establishing its existence, and some of its properties, prior to its being observed. Their feat is equally surprising whether we say Neptune was at the time unobservable, or observable but not yet observed.

[4]One representative of this type of view was Henri Poincaré. He held that, although principles such as the conservation of mass are *suggested* by experiment, the law that, for instance, mass is conserved implicitly defines our notion of mass. In his view, although the law of conservation of mass could be rejected because it was no longer useful, it could not be falsified.

[5]Drell also mentions impressive confirmations for QED with respect to the predicted values of the 'first excited states of positronium', the 'helium fine structure', 'muonic X-rays' and the Lamb shift in μ-helium. See Drell (1979, pp. 12-13).

[6]The value of the magnetic moment of the electron obtained by experiment is $1159652410 \times 10^{-12}$ while the value predicted by QED is $1159652359 \times 10^{-12}$. If we subtract the second figure from the first, the result is 51×10^{-12}. What are the chances of obtaining this close a degree of agreement between theory and observation? This will depend upon the number of possible observations that *could have* been made. From a purely *a priori* point of view, there is an indefinitely large number of results that could have been obtained: the results could have ranged anywhere from zero upwards. It would, of course, be a mistake to suppose that the finite capacity of the measuring apparatus, or the finite time we have to record results, should place any restriction on the range of possible results. The outcome of the experiment might have been a magnitude too large for us to measure, or the apparatus

might have melted, or exploded, or something quite bizarre could have happened. It would seem all such outcomes ought to receive some non-zero, although very small, probability. Therefore, we are on safe ground if we say that the range of *possible* results goes from zero to *at least* $2{,}000{,}000{,}000 \times 10^{-12}$. The result that was obtained by experiment was within $\pm 51 \times 10^{-12}$ of the predicted result, or within a region of a size of about 100×10^{-12}. So, it seems reasonable to assert that the *a priori* probability of obtaining a result in this region is not more than $100 \div 2{,}000{,}000{,}000$; or 5×10^{-8}. This is an *over*estimation of the probability, since it is based on the assumption that the probability of any result greater than $2{,}000{,}000{,}000 \times 10^{-12}$ is zero, and there is no *a priori* reason why that should be the case.

[7] Perhaps the best known exponent of this view is Richard Boyd. See Boyd (1983).

[8] I do not know of anyone who has explicitly developed a meta-inductive explanation of the success of science. However, W.H. Newton-Smith (1981, especially Chapter 9) has argued that scientific method grows in much the same way as the content of science itself grows.

[9] I owe this objection to Robert Nola.

[10] One philosopher who has offered a broadly 'evolutionary', or trial and error, account of the success of science is Larry Laudan. See, for example, his (1984, especially p. 101). Laudan also attaches great importance to the use of double-blind testing in science. But plainly, this does not explain *novel* success.

[11] One prominent scientist who has suggested this as an explanation of the ability of science to make successful predictions in novel domains is Richard Feynman (1992, p. 173).

[12] This suggestion was put to me, in conversation, by Peter Lipton.

LISA BORTOLOTTI

MARKS OF IRRATIONALITY

1. INTRODUCTION

Philosophers' attention has recently been drawn to the phenomenon of delusions with clear organic causes. Delusions are usually defined as beliefs with very implausible content that are maintained in the face of strong counterevidence.[1] A satisfactory analysis of delusions must account for their being irrational beliefs and this raises interesting philosophical issues. First, the same phrase 'irrational beliefs' is considered by some to be paradoxical, or at least very problematic.[2] If, in order to ascribe beliefs to a system *S*, we have to assume that *S* is rational, then how can we ascribe *irrational* beliefs to *S*? And, even if we are indifferent to the concerns derived from the rationality constraint on belief ascription, delusions remain a puzzle. Are they genuine beliefs? Are they really irrational? If so, how do they differ from other irrational beliefs?

These are all extremely controversial issues, and this paper will only partially address them. The object of the paper will be to characterise the sense in which delusions can be said to be irrational beliefs. Here I shall not defend the claim that delusions are genuine beliefs, but I shall draw attention to the fact that delusions are at least functionally very similar to belief states. At least some delusions participate in inferences, are argued for, and held with sincere conviction and they sometimes lead to action.

The irrationality of delusion will be my main focus. My suggestion will be that a belief system affected by delusions is irrational primarily because the delusional beliefs resist revision in circumstances in which (i) they conflict with other beliefs in the system and (ii) independent evidence against them becomes available. The possible tension between the subject's delusional beliefs and her other beliefs undermines the coherence of the system, and the obstinacy with which the subject holds on to her delusional beliefs compromises the sensitivity of the system to new evidence.

S. Clarke and T.D. Lyons (eds.), Recent Themes in the Philosophy of Science, 157–173.

As a preliminary, I shall argue that we do not need to worry about the presumed paradoxical nature of irrational beliefs. There seems to be nothing incoherent in the ascription of beliefs that are less than rational. Then, I shall describe a case of delusion, the Capgras syndrome, and argue that at least some cases of Capgras delusion can be described in terms of irrational beliefs. Then, I shall point out that the features of delusions, which I have introduced as markers of irrationality, *resistance to change* and *compartmentalisation*, can also be shared by non-delusional beliefs.

This observation will prompt some further reflection on the nature of delusions. Those philosophers who want to deny that delusions are beliefs cannot appeal only to the fact that delusions are resistant to change and are compartmentalised. In the psychiatric literature there are independent arguments against the definition of delusions as beliefs, but such arguments will not be discussed here.[3] What I shall not deny is that resistance to change and compartmentalisation can significantly contribute to the irrationality of both delusions and non-delusional beliefs. Indeed, the distinctiveness of delusional beliefs will partially evaporate, as I argue that delusions and non-delusional beliefs, when irrational, are irrational for the same reasons. Important questions about the nature of beliefs will emerge, as we are in need of some criteria to distinguish delusions from other irrational beliefs. Were epistemology to fail in characterising why delusions are special, the appeal to their organic causes might turn out to be their only distinctive feature.

2. RATIONALITY AND INTENTIONALITY

In this section, I will briefly explain why we should not worry about the phrase 'irrational beliefs' being incoherent or paradoxical. For the last thirty years, Donald Davidson has argued that we need to assume that a system is largely rational in order to describe it as an intentional system. The rationality constraint on belief ascription is based on the idea that there is a necessary link between rationality and intentionality. Where there is no rationality, there can be no intentional description and, therefore, no description of mental states in terms of beliefs.

The rationality constraint has been challenged in several contexts. An objection to the alleged necessary link between rationality and intentionality comes from the observation that we do ascribe beliefs to people who make systematic reasoning mistakes, are inconsistent and hold absurd beliefs. If we combine the theory of belief ascription based on the rationality constraint with the experimental data on the limitations and biases of human reasoning[4], then we apparently have to conclude that we are almost never genuine believers. However, we do ascribe beliefs on an everyday basis, for instance when we engage in conversation, and we often ascribe beliefs not just to other speakers, but also to infants and to some non-human animals. It has been argued that our practice of everyday belief attributors is at odds with the idea that there is a necessary rationality assumption embedded in our practices of intentional explanation and prediction. We do not seem to assume that only a rational creature can be ascribed intentional states. In fact, adults are seldom fully rational, and infants and non-human animals are probably never so, but we characterise them as believers, at least in some circumstances.

In an attempt to reply to this general objection, Davidson has offered a slightly revised version of his account of belief ascription stressing that the constraint does not rely on the demanding notion of perfect rationality. He claims that not all forms of irrationality are necessarily banned from intentional systems. Local deviations from norms of truth, consistency or good reasoning are to be viewed as exceptions that can be explained only against a wider background of rationality (Davidson 1985, pp. 345-354).

> Background Argument: the ascription of non-rational beliefs to system *S* is not incompatible with the rationality constraint on belief ascription, because all the rationality constraint was ever meant to say was that *S* must be *largely* rational in order to qualify as an intentional system.

What is this 'background' of rationality? In a system that is largely rational, beliefs are formed in a reliable way, are generally true, cohere with each other, and might be rejected or revised when a tension emerges among them, or when new conflicting data becomes available. In Davidson's terminology, the system is largely rational if the 'logical relations' among beliefs and their 'causal relations to events and objects in the world' are in good order.

The background argument offers a plausible view of cognitive rationality, but it is not sufficient to defend the necessary link between rationality and intentionality from further challenges. For instance, can systems still be intentional systems, if affected by delusions?

> We have no trouble understanding small perturbations against a background with which we are largely in sympathy, but large deviations from reality or consistency begin to undermine our ability to describe and explain what is going on in mental terms.
>
> (Davidson 1982, p. 303)

It is open to the Davidsonian to claim that delusions can be described in intentional terms, if we apply the background argument to them. On what grounds can systems affected by delusions be considered as largely rational? One could convincingly argue that delusions are *localised* irrationalities against a background of rationality, and should not be treated differently from reasoning mistakes and other patently false or inconsistent beliefs.

Three factors seem to suggest that this is a viable analysis. First, a delusional state might not affect the entire belief system. Consider the case in which the subject has a *monothematic non-elaborated* delusion, that is, a delusion that is confined to one specific topic and the patient does not act in accordance with it nor makes inferences from it. In this case, it is possible that the subject's other beliefs turn out to be perfectly rational. The second factor is the apparent inclusion of the belief in a rational pattern. When asked to justify her delusional state, the subject might appreciate the implausibility of the content of her delusion, showing that she is still able, to some extent, to weigh up reasons for her beliefs and recognise that her views are not likely to be shared. Finally, some patients recover and give up their delusional states. These phenomena might speak in favour of the application of the background argument to the case of delusions.

However, there are two main reasons to be cautious about the extension of the background argument to the case of delusions in general. The Davidsonian account cannot easily support the analysis of *polythematic elaborated* delusions as temporary and localised deviations from rationality. It would be implausible to apply the background argument to an entire *delusional system*, that is, a system where the delusional states are not confined to one or two thematic areas, but widely influence most of the patient's beliefs and actions. In some cases of schizophrenia, the patient develops a narrative and tends to interpret virtually all events in her life in the light of her delusions.[5]

We might also deny that the background argument can successfully account for the intentional description of non-elaborated delusional states, in spite of the previous considerations about insulation, rational patterns and recovery. Implicit in the background argument is the idea that the presence of a non-rational belief can only be explained by appealing to the other things that the subject rationally and truly believes.

> [...] When a mistake is agreed to have been made we will often look for, and find, a reason why it was made, not just in the sense of cause or regularity in its making but in the sense of some excuse which reconciles the mistake with the idea that, even in making it, the perpetrator was exercising his or her rationality. (Heal 1998 p. 99)

The attempt to rationalise an implausible belief by reference to other beliefs the subject holds is often successful. The five year-old who believes that Santa Claus left three presents for her under the Christmas tree, has a false belief by our standards. However, her belief can be easily rationalised by appealing to another true belief of hers, that her mother is generally a reliable source of information. If her mother told her that Santa Claus would have left the presents under the tree, then the five year old has an apparently good reason to believe that the presents were brought by Santa Claus.

This rationalisation manoeuvre seems not to be available for patients who suffer from monothematic and relatively circumscribed delusions, as it will become clear in the next sections when we look at some concrete examples. In such cases, there seems to be no independent belief in the patient's belief systems to which she can appeal in order to rationalise the acceptance of the delusion. Only the subject's confidence in the content of her own experience grounds her belief. What we can do as interpreters is to explain the formation of the delusional state by reference to its organic causes, namely the patient's brain damage. From a Davidsonian perspective, there are no supporting beliefs that justify the subject's holding on to her delusional state, either from the point of view of the subject herself, or from that of a charitable interpreter.

> In standard reason explanation [...] not only do the propositional contents of various beliefs and desires bear appropriate logical relations to one another and to the contents of the belief, attitude or intention they help to explain; the actual states of belief and desire cause the explained state or event. In the case of irrationality, the causal relation remains, while the logical relation is missing or distorted. [...] There is a mental cause that is not a reason for what it causes. (Davidson 1982, p. 298)

This is not the place to assess the adequacy of the rationality constraint, but it should be clear that belief ascription theories that rely on it seem to lack the resources to account for the intentional characterisation of delusional states. One easy way to maintain common sense characterisations is to deny that we need a rationality constraint on belief ascription. We can retain the rationality assumption merely as a useful and fallible heuristic in dealing with human believers.

Whether you share my scepticism about the necessity of the rationality constraint, or are persuaded by the extension of Davidson's background argument to the case of delusions in spite of my reservations, the idea that the notion of 'irrational belief' is paradoxical should by now have lost its appeal. In what follows, then, I shall assume that there is nothing incoherent in the idea that delusions are irrational beliefs.

3. THE CAPGRAS SYNDROME

In this section I will briefly describe the Capgras delusion. Patients affected by the Capgras syndrome claim that their spouses, or very close relatives, have been replaced by impostors. According to a widely accepted view,[6] the delusion arises when the affective component of the face processing module is damaged, leaving recognition unimpaired. The patient sees the spouse's face and recognises it, but does not experience any affective response. On one possible interpretation, the delusion is an attempt to explain why the face seen, which appears identical to the familiar face, 'feels' strange (Gerrans 2000, pp. 111-122). On most accounts, a reasoning bias (or a reasoning deficit) is also necessary for the formation of this delusion.[7] Further studies reveal that there is also an auditory form of the Capgras delusion, where the recognition of familiar voices is affected. This auditory impairment seems to be compatible with the hypothesis that the Capgras delusion is correlated with damages to the affective channel of recognition mechanisms.

Stone and Young (1997) report the case of a 44 year-old man who, after brain damage due to a traffic accident, started claiming that his family and house had been duplicated. Young (2000) refers to the case of ML, a 60 year-old woman also affected by Capgras delusion, probably due to cerebral infraction, who thought that her son was an impostor and that he wanted to kill her. Reid et al. (1993, pp. 225-228) describe the puzzling case of a 32-year-old *blind* woman who experienced the Capgras delusion. She believed that her pet cat had been replaced by a replica that was hostile to her.

In most common instances, the Capgras syndrome is a monothematic delusion, that is, it is often confined to the presumed substitution of the spouse or close relative with a clone, an alien or a robot that looks identical (or almost identical) to the original person. In some cases, Capgras patients act on their delusional beliefs by showing hostile or aggressive behaviour towards the presumed impostors. The character of delusions as action guiding seems to be occasionally supported by the tragic stories of some patients. One case often reported is that of Ms. A. who killed her mother, after having suffered for five years the delusion that her parents were impostors (Silva et al. 1994, pp. 215-219).

Pressed by questions, Capgras patients can also form other beliefs related to the content of their delusional state in the attempt to explain why their delusion fits so badly with other things they know. A patient could be asked to explain why the 'impostor' has the ring he gave to his wife, and the patient could reply that the ring is not the same one, just a very similar one. When patients are asked why they had not reported the disappearance of their spouses to the police, they often candidly answer that the police would have never believed them. The more complex and articulated the explanations given by the patients are, the more elaborated is their delusional state, where elaboration is measured in terms of the connections that the delusional state has with other beliefs the patients hold. Alternatively, patients can go on with their lives substantially unchanged and show very little concern about their unusual situation and the presumed disappearance of their spouses or close relatives. In these latter cases the delusion is said to be circumscribed, that is, the delusional state does not necessarily interact with the patient's other beliefs and does not necessarily lead the patient to act in accordance with it, probably as an effect of some form of compartmentalisation.

A general feature of both elaborated and circumscribed delusions seems to be the obstinacy with which the patient maintains the delusion, in spite of its having a very implausible content, and in spite of other people's attempts to dissuade her. The fact that the delusional state is preserved in the face of strong evidence and testimony against it has been highlighted in order to argue that the way delusions work is significantly different from the way beliefs do. I challenge that claim by showing that some representational states that we regard as uncontroversial instances of belief can also be compartmentalised and resistant to change. Nevertheless, the extent to which resistance to change and compartmentalisation affect the patient's ability to evaluate and ultimately reject her delusional state, indicates that there is an element of irrationality in the subject's commitment to the content of her delusion.

While it seems uncontroversial that delusions are somehow a manifestation of irrationality, theorists disagree about the sense in which delusions can be said to be irrational. What characteristics of delusions are uncontroversial indications of irrationality? The diversity of opinions also reflects the general confusion that surrounds theoretical notions of rationality. My suggestion is that the belief systems might fail to be rational for one or more of the following reasons:

1. A very implausible belief is formed without there being sufficient or adequate justification;

2. A belief is maintained in the face of strong counterevidence;

3. A belief might be compartmentalised – that is, it might not cohere with other beliefs that belong to the same system.

These three possible failures of rationality correspond to three constraints that characterise any belief system: formation, revision and integration of beliefs. Mine is not an entirely original suggestion. Gerrans (2000) and Young (2000) also identify the irrationality of delusions with failures at the level of coherence and revisability.

> Rationality is a normative constraint of consistency and coherence on the formation of a set of beliefs and thus is *prima facie* violated in two ways by the delusional subject. First she accepts a belief that is incoherent with the rest of her beliefs, and secondly she refuses to modify that belief in the face of fairly conclusive counter-evidence and a set of background beliefs that contradict the delusional belief (Gerrans 2000, p. 114).

> Delusions do not seem to respect the idea that the belief system forms a coherent whole and that adjustments to one belief will require adjustments to many others.
>
> (Young 2000, p. 49)

Ideally, we would want our belief systems to include mostly well-justified or justifiable beliefs, to be sensitive to counterevidence and to exclude beliefs that have been disconfirmed or are in tension with other justified beliefs that are already included in the system. The 'epistemic virtues' I have in mind are justification, responsiveness to evidence and coherence. This is by no means supposed to be an exhaustive account of the rationality of beliefs, but rather a rough guide to the detection and classification of deviations from rationality. How do delusions fare in relation to my three criteria?

Maher has argued that delusions are not ill-formed beliefs. He believes that there is nothing wrong in delusional subjects' reasoning. The abnormality of the delusional beliefs would be entirely due to the abnormality of the experiences on the basis of which the subjects have formed their beliefs. The idea is that delusional subjects do not violate normative standards of reasoning more often than other believers do. The only real difference is that delusional subjects have to account for certain abnormal experiences caused by brain damage. The content of those experiences is responsible for the implausible content of their delusional beliefs (Maher 1974, pp. 98-113).

Maher's suggestion is an interesting one, but cannot have the consequence of establishing that delusions are not irrational. Even were we to deny that belief formation is affected in delusional subjects, we would still need to account for their compartmentalisation and resistance to counterevidence. Let us assume that the belief formation process is intact. It does not follow from that that Capgras patients are rational believers and that the abnormality of their delusional beliefs depends exclusively on the abnormal experiences from which they originate. One way of making this point is to refer to the case of everyday perceptual illusions. We see the straw in the glass of water as bent or broken. We might even come to believe — for instance the first time that we are subject to the illusion — that the straw *is* bent. But then we revise our belief when we realise that our visual experience was misleading. Capgras patients have many good reasons to doubt the content of their experience, including the authority of their psychiatrists and the testimony of their relatives. More relevantly to the analogy with perceptual illusions, the sensation of unfamiliarity often evaporates when the patient talks to the 'impostor' on the phone, in the same way as the content of our experience changes when we view the straw out of the water. Experience often misleads us, but we do not maintain the belief we have formed as a consequence of our having a certain experience, if the belief strikes us as implausible given other things we know, or some other experience. We suspend judgement or just revise the belief. In short, rationality does not concern belief formation alone.[8]

> Regardless of whether the delusional beliefs spring from bizarre perceptual experiences, it is far from clear that we ought to jettison the criterion of imperviousness to counterevidence or that we ought to allow that delusional beliefs are good explanations. (Leeser and O'Donohue 1999, p. 690)

4. REVISION

In a typical scenario, the Capgras patient comes to believe that, say, her mother has been replaced by an impostor. The patient holds on to her belief, even though the 'duplicate' acts and looks exactly like her mother, has her mother's memories and is recognised as her mother by other relatives and friends. Moreover, a figure of authority, typically a psychiatrist, informs the patient that she is suffering from a well-known syndrome due to brain damage. Still, the patient's belief that her mother has been replaced by an impostor is unshaken.

Delusional states resist revision in an anomalous way and have been defined as beliefs held in the face of evidence normally sufficient to destroy it. The dynamic nature of beliefs is taken very seriously, and some authors come to identify the distinctive feature of beliefs with their capacity to change:

> Belief [...] is not solely a disposition to behave in certain ways should certain situations arise, nor is it simply an enduring mental state of some kind, and scarcely a mere stored representation. It is a live entity, drawing upon the environment and reacting with it, in other words, a dynamic and functional phenomenon. (Jones 1999, p. 3)

If the dynamic nature of beliefs is what distinguishes them from other mental representations, as Jones seems to think, then there is a point in denying delusions the status of beliefs. Nevertheless, resistance to change is not an exclusive character of delusions and is not necessarily a manifestation of irrationality. Resistance to change is an indication of irrationality when the belief content remains unaltered, even though the reasons why the belief was accepted cease to provide adequate justification for it and new evidence against that belief is available.

In this section I shall offer some examples of non-delusional beliefs that are resistant to change. My goal will be to show that the perseverance that characterises delusional subjects is not uncommon among believers in general. There have been many psychological studies on belief perseverance. In particular, resistance to change has been studied in relation to theory dynamics within the history of science. Epistemologists have thoroughly discussed the importance of the principle of conservatism and its consequences for the structure of belief systems. In every belief system there is 'a tension between forming beliefs that require little readjustment to the web of belief (conservatism) and forming beliefs that do justice to the deliverances of one's perceptual systems' (Stone and Young 1997, p. 349). The conflict between consistency and conservatism, on the one hand, and observational adequacy and revisability, on the other, is a central issue in the study of cognitive rationality. The following cases of resistance to change raise the same issue.

(*a*) *The earth had existed long before my birth* (Wittgenstein 1969, §84 & §92).

What is special about (*a*)?

> What we call historical evidence points to the existence of the earth a long time before my birth; — the opposite hypothesis has *nothing* on its side (Wittgenstein 1969, §190).

Proposition (*a*) is a proposition on which our knowledge of, say, geography, history and astronomy is based. It is also part of our common sense, as we soon recognise that our parents look older than we are and seem to have spent their lives on this planet before our births. If we rejected (*a*), we would need to reject many other beliefs that depend on (*a*), and probably change our entire world view. That the earth had existed long before our births is a *hinge* proposition. That is why we would resist any attempt to persuade us that the earth came into existence just a few minutes before we did. In cases like (*a*) the phenomenon of resistance to change is not a manifestation of irrationality, but an indication of the special role that the proposition plays in our cognitive life and an appreciation of the fact that we are not clear about what *kind* of evidence would count against it. Fiction might help us imagine scenarios in which we could come to reject (*a*), but there is a strong prima facie doubt whether those scenarios would be coherent.

The second case I would like to consider is a case of self-deception.

(*b*) *My wife is faithful.*

> Suppose that Sam has believed for many years that his wife Sally would never have an affair. In the past, his evidence for this belief was quite good. Sally obviously adored him; she never displayed a sexual interest in another man; she condemned extramarital sexual activity; she was secure and happy with her family life; and so on. However, things recently began to change significantly. Sally is now arriving home late from work on the average of two nights a week; she frequently finds excuses to leave the house alone after

> dinner; and Sam has been informed by a close friend that Sally has been seen in the company of a certain Mr. Jones at a theatre and local lounge. Nevertheless, Sam continues to believe that Sally would never have an affair. Unfortunately, he is wrong. Her relationship with Jones is by no means platonic. (Mele 1987, pp. 131-2)

Although the acceptance of the fact that his wife is having an affair would not shake Sam's entire world view, it would have a very strong impact on his feelings, and cause him great distress. This is a common phenomenon with beliefs that are emotionally charged. Giving up (*b*) can become very painful for Sam.

> When [...] a clear and unequivocally disconfirming evidence impinges on a person, the cognition corresponding to this knowledge is dissonant with the belief he holds. When such a state of affairs exists, the most usual and ordinary way of eliminating the dissonance is to discard the belief rather than attempt to deny the evidence of one's own senses. [...] But there are circumstances in which this does not happen – that is, even in the face of clearly disconfirming evidence, the belief is not discarded.
>
> (Festinger 1957, p. 244)

Taking on board the evidence against (*b*) would be dissonant with Sam's desire to believe his wife trustworthy. Such dissonance may delay the rejection of Sam's belief in his partner's faithfulness, or, in a clearly self-deceptive case, make it impossible. According to Mele (1987), Sam adopts a self-deceptive strategy that consists in misinterpreting the evidence, focusing selectively on the facts that support his belief that Sally is faithful and gathering new data in an equally selective way, by avoiding those data that would undermine his belief. Sam's belief that Sally is faithful becomes unresponsive to the evidence available to him. Sam's belief in (*b*) is firmly resistant to change, and can be described as irrational or pathological, but it is not obviously delusional. In fact, it does not meet one of the criteria for delusional beliefs, a wildly implausible content.

Let us consider now the historical case of a firm belief in a scientific hypothesis:

(c) *Phlogiston explains combustion.*

The scientist who tests a hypothesis assumes it is true and attempts to accommodate evidence in order to explain away the discrepancies between the predictions made in accordance with the hypothesis and the actual outcome of the tests. There could be good reasons why the scientist does not immediately take the unexpected results as challenging the truth of her hypothesis. But, at least according to a positivist methodology, if the evidence against the hypothesis accumulates and better explanations of the new evidence are available, then the scientist should give up her initial hypothesis. There is one famous case in which this did not happen, and I shall mention it here as an instance of resistance to change in science.

The chemical revolution originated with a competition between two rival theories that both accounted for the same experimental results and could both explain a wide range of phenomena. Eventually, the old theory (phlogiston theory) was overthrown and replaced by a new one (oxygen theory), whose satisfactory formulation was articulated only after Lavoisier suggested that oxygen was the purest portion of the air and that its presence was necessary for combustion. The theoretical shift was rather dramatic, but also spread over a long period of time, and for this reason many scientists at different times and on several occasions contributed to it. In the end, the phlogiston theory was completely rejected and replaced by the oxygen theory thanks to Lavoisier's theoretical innovations. All the scientific community of chemists working in that area (with the noteworthy exception of Joseph Priestley) adopted the new model and continued their research in the light of it.

Priestley, who was committed to the phlogiston theory, did not ignore the anomalies Lavoisier tried to solve with his oxygen theory. On the contrary, he tried to solve them himself, by making revisions to the phlogiston theory, to which he was committed. When he isolated 'dephlogisticated air' (what we call 'oxygen' today), he reserved an explanatory role for it in his system. He also drew a connection between the purity of the new gas and its effects on combustion. Until his death, by which time Lavoisier's theory had already been widely accepted in the scientific community, Priestley still maintained that 'his' phlogiston theory was true. Many philosophers and historians, including Kuhn and Lakatos, have commented on Priestley's obstinacy.

> Though the historian can always find men who were unreasonable to resist for as long as they did, he will not find a point at which resistance becomes illogical or unscientific.
> (Kuhn 1970, p. 159)

Resistance to change is not seen by Kuhn as an instance of irrationality, because it is closely related to what he considers a constitutive notion for science, the (perhaps obstinate) adhesion to a paradigm within which scientists have been trained and have been working. Defending one's belief against opposition is no less rational than being ready to admit one's mistake. Ayer vividly describes the tension between dynamic and conservative forces in the making of science.

> It is disagreeable and troublesome for us to admit that our existing system is radically defective. And it is true that, other things being equal, we prefer simple to complex hypotheses, again from the desire to save ourselves trouble. But if experience leads us to

> suppose that radical changes are necessary, then we are prepared to make them, even though they do complicate our system, as the recent history of physics shows. When an observation runs counter to our most confident expectations, the easiest course is to ignore it, or at any rate explain it away. (Ayer 1946, pp. 98-99)

My suggestion in this section has been that resistance to change does not only characterise delusional beliefs and is not necessarily an indication of irrationality. But what about compartmentalisation?

5. INTEGRATION

Let us go back to the case of the Capgras patient who believes that her mother has been replaced by an impostor. Why do we say that her belief is compartmentalised? One reason is that the patient recognises the implausibility of the idea that, say, aliens have come to the earth to kidnap her mother, duplicate her and replace her with an alien in disguise. Yet, these considerations about the low credibility of the patient's hypothesis do not affect her certainty in the belief that her mother has been duplicated.

Since beliefs usually cohere and support each other in a system, the fact that delusions might not significantly affect the subject's other beliefs (and might not dispose the subject to act on them) has been offered as a reason for the claim that delusions are not genuine beliefs. Some reflection will help us realise that there are representational states that we are happy to classify as beliefs even though they are not well integrated in the system of beliefs we have. We do not act on each of our beliefs, and some of our beliefs do not have a great impact on the rest of our cognitive life. They just 'sit there'.

Let us focus on two examples.

(*d*) *One should have safe sex.*

A few years ago, some studies were conducted in the US about the use of condoms by college students. It soon emerged from interviews and surveys that college students were definitely in favour of the use of condoms for the prevention of AIDS. Nonetheless, evidence from other sources indicated that college students did not use condoms regularly, mainly because they considered it 'a nuisance'. As Aronson put it, they seemed to 'deny that the dangers of unprotected sex applied to them in the same way as they applied to everyone else' (Aronson 1999, p. 114). If the evidence collected by the psychologists is reliable, then the students' belief that one should not have unprotected sex is compartmentalised and does not play a role in guiding their own behaviour. It is not even clear from their reports that avoiding AIDS is a consideration for them to weigh in the process of making a decision about whether to use condoms. If the importance of preventing sexually transmitted diseases crosses their minds at all, it is then outweighed by the belief that condoms are a nuisance. However, a better explanation is that the belief (*d*) does not even interact with the other beliefs about condom use and that this is a case of poor integration among beliefs. Aronson points out that this phenomenon is not at all an exception. Often we sincerely endorse general principles that we fail to apply in the appropriate context.

(*e*) *It is appalling that there are still people who starve to death in developing countries.*

Many of us would claim to be appalled by the living conditions of the inhabitants of developing countries. Nonetheless, very few of us do anything about it. We might show no interest about a recent emergency in India. We might avoid talking about the problems in developing countries with our friends or give no support to development projects. Our belief that nobody in the 21st century should die of starvation might not interact with other beliefs we hold, for instance, our belief about what interesting topics of conversation there are, or how we should spend our time and money. We might be regarded as incoherent, but our belief that everybody should have decent living conditions is no less a belief just because it is scarcely manifest in the way we behave.

It seems very plausible to say that, in all belief systems there are compartmentalised beliefs, and the partitioning of our beliefs in subsystems has often been advocated in order to account for everyday deficits of rationality. The fact that we sometimes hold inconsistent beliefs, that we are incontinent, deceive ourselves or reveal hypocrisy can be explained by appealing to some form of compartmentalisation.[9]

6. CONCLUSION

In this paper I have explored the issue of whether delusions are irrational beliefs. As a preliminary move, I have argued that there is nothing incoherent about the idea of beliefs that are irrational, and that therefore the characterisation of delusions as irrational beliefs is a viable one. But is there any good reason to support this characterisation? I have suggested that the answer to the question of whether or not delusions are irrational depends upon (1) the way in which delusions are formed, (2) the circumstances in which they are (or are not) revised and (3) whether they cohere with other beliefs the subject holds. It seems that we can identify some instances of resistance to counterevidence and compartmentalisation of delusional beliefs as marks of irrationality.

At this point, just when we seem to have singled out what is so puzzling about delusions, new worries emerge. Non-delusional beliefs can also be compartmentalised and resistant to counterevidence. Recall the rough definition of delusions I started with in my introduction. Delusions are beliefs with very implausible content that are maintained in the face of strong counterevidence. This 'rough definition', together with more respectable ones, fails to provide a satisfactory account of what distinguishes delusions from non-delusional beliefs that exhibit marks of irrationality. Unless we focus on the specific contents of delusional beliefs, or on the brain damage that is the organic cause of the formation of the delusions, we cannot easily spell out the difference. By tracing resistance to change and compartmentalisation, in uncontroversial cases of belief, the epistemic divide between delusional subjects and other believers has been narrowed.[10]

Australian National University

REFERENCES

Aronson, E.: 1999, 'Dissonance, Hypocrisy and the Self-Concept', in E. Harmon-Jones and J. Mills (eds.), *Cognitive Dissonance*, American Psychological Association, Washington D.C., pp. 103-126.

Ayer, A.: 1946, *Language, Truth and Logic*, Victor Gollancz, London.

Baker, C. and Morrison, A.: 1998, 'Cognitive Processes in Auditory Hallucinations: Attributional Biases and Metacognition', *Psychological Medicine* 28, 1199-1208.

Berrios, G.: 1991, 'Delusions as 'Wrong Beliefs': A Conceptual History', *British Journal of Psychiatry* 159, suppl. 14, 6-13.

Breen, N., Caine, D. et al.: 2000, 'Towards an Understanding of Delusions of Misidentification: Four Case Studies', *Mind & Language* 15(1), 74-110.

Campbell, J.: 1999, 'Schizophrenia, the Space of Reasons, and Thinking as a Motor Process', *The Monist* 82(4), 609-625.

Cherniak, C.: 1986, *Minimal Rationality*, MIT Press, Cambridge (Mass.).

Davidson, D.: 1982, 'Paradoxes of Irrationality' in R. Wollheim (ed.), *Philosophical Essays on Freud*, Cambridge University Press, London, pp. 289-305.

Davidson, D.: 1985, 'Incoherence and Irrationality', *Dialectica* 39(4), 345-354.

Davies, M. and Coltheart, M.: 2000, 'Introduction: Pathologies of Belief', *Mind & Language* 15(1), 1-46.

Dennett, D.: 1987, *The Intentional Stance*, MIT Press, Cambridge (Mass.).

Festinger, L.: 1957, *A Theory of Cognitive Dissonance*, Stanford University Press, Stanford.

Garety, P.: 1991, 'Reasoning and Delusions', *British Journal of Psychiatry* 159, suppl. 14, 14-18.

Gerrans, P.: 2000, 'Refining the Explanation of Cotard's Delusion', *Mind & Language* 15(1), 111-122.

Heal, J.: 1998, 'Understanding Other Minds from the Inside' in A. O'Hear (ed.), *Current Issues in Philosophy of Mind*, Cambridge University Press, Cambridge, pp. 83-100.

Hemsley, D. and Garety, P.: 1986, 'The Formation and Maintenance of Delusions: a Bayesian Analysis', *British Journal of Psychiatry* 149, 51-56.

Jones, E.: 1999, 'The Phenomenology of Abnormal Belief: A Philosophical and Psychiatric Inquiry', *Philosophy, Psychiatry, and Psychology* 6(1), 1-16.

Kahneman, D., Tversky, A. and Slovic, P.: 1982, *Judgement Under Uncertainty: Heuristics and Biases*, Cambridge University Press, Cambridge.

Kuhn, T.: 1970, *The Structure of Scientific Revolutions*, University of Chicago Press, Chicago.

Langdon, R. and Coltheart, M.: 2000, 'The Cognitive Neuropsychology of Delusions', *Mind & Language* 15(1), 184-218.

Leeser, J. and O'Donohue, W.: 1999, 'What Is a Delusion? Epistemological Dimensions', *Journal of Abnormal Psychology* 108(4), 687-694.

Maher, B.: 1974, 'Delusional Thinking and Perceptual Disorder', *Journal of Individual Psychology* 30, 98-113.

Maher, B.: 1999, 'Anomalous Experience in Everyday Life: Its Significance for Psychopathology', *The Monist* 82(4), 547-570.

Mele, A.: 1987, Irrationality. *An Essay on Akrasia, Self-deception and Self-control*, Oxford University Press, Oxford.

Nisbett, R. and Ross, L.: 1980, *Human Inference: Strategies and Shortcomings of Social Judgment*, Prentice-hall, Englewood Cliffs N.J.

Payne, R.: 1992, 'My Schizophrenia.', Schizophrenia Bulletin 18(4), 725-728.

Reid, I., Young, A. et al.: 1993, 'Voice Recognition Impairment in a Blind Capgras Patient', *Behavioural Neurology* 6(4), 225-228.

Rust, J.: 1990, 'Delusions, Irrationality and Cognitive Science', *Philosophical Psychology* 3(1), 123-137.

Sass, L.: 1994, *The Paradoxes of Delusion*, Cornell University Press, Ithaca and London.

Silva, J., Leong, G. et al.: 1994, 'Delusional Misidentification Syndromes and Dangerousness', *Psychopathology* 27, 215-219.

Stone, T. and Young, A.: 1997, 'Delusions and Brain Injury: The Philosophy and Psychology of Belief', *Mind & Language* 12, 327-364.

Wittgenstein, L.: 1969, *On Certainty*, Blackwell, Oxford.

Young, A.: 2000, 'Wondrous Strange: The Neuropsychology of Abnormal Beliefs', *Mind & Language* 15(1), 47-73.

NOTES

[1]'Psychiatric diagnostic manuals usually define delusions in terms of a combination of their implausibility to other people with a similar cultural background and their resistance to counter-suggestion or obvious evidence to the contrary' (Young 2000, p. 47). According to the Diagnostic and Statistical Manual of Mental Disorders (DSM-IV 1994, p. 765), a delusion is 'a false belief based on incorrect inference about external reality that is firmly sustained despite what almost everyone else believes and despite what constitutes incontrovertible and obvious proof or evidence to the contrary'.

[2]According to Davidson 'The idea of an irrational action, belief, intention, inference or emotion is paradoxical' (1982, p. 289).

[3]Some deny that delusions are belief states. Berrios (1991) believes that delusions are empty speech acts misidentified by the patients as beliefs. Sass (1994) claims that the content of delusions is meant to describe the private world of the patient's experience and not the real world.
[4]See Kahneman et al. (1982), Nisbett and Ross (1980) and Cherniak (1986).
[5]See Payne (1992). After a history of depression and alcoholism, Roberta started having vivid visual hallucinations and experiencing 'TV broadcasting'. Other hallucinations and delusions followed, including people's substitution, alien control and persecution. Though implausible, and sometimes not clearly related to one another, Roberta's delusional beliefs are part of a general narrative. It seems as if she interprets all the events happening in her life in the light of her beliefs about the power of the alien beings and the hostility of other people. Roberta's most important decisions, whether to keep in touch with her family, to change job and to move to another city, were made on the basis of her delusional beliefs and the fear and anxiety that they were causing her.
[6]See Stone et al. (1997, pp. 327-364).
[7]See Langdon et al. (2000, pp. 184-218) and Davies et al. (2000, pp. 1-46).
[8]According to Langdon: 'If the presence of a perceptual aberration were a sufficient condition for the presence of a delusional belief, then any individual who had an aberrant perceptual experience would develop a delusional belief. But that is not so' (Langdon 2000, p. 190).
[9]According to Davidson: '[…] If we are going to explain irrationality at all, it seems we must assume that the mind can be partitioned into quasi-independent structures […] (Davidson 1982, p. 300).
[10]Thanks to Martin Davies, Kim Sterelny, Peter Godfrey-Smith and Matteo Mameli for stimulating discussion on the topic of this paper. Thanks also to Laura Schroeter and Daniel Cohen for very helpful comments on the final draft.

HERMAN C.D.G. De REGT

IMPLICATIONS OF INQUIRY: A PRAGMATIST VIEW OF THE SCIENTIFIC AND MANIFEST IMAGE

1. INTRODUCTION

Normally, one associates 'the manifest image' with the perspective of the unreflective common man and 'the scientific image' with the perspective of natural science. It is worthwhile reconsidering this dichotomy of images because it suggests the radically misleading idea that the world of natural science is more basic than the ordinary and everyday world of the common man. Natural science depicts, or will one day depict, the *true* ontology of the world, thereby replacing, reducing or flat out eliminating common sense notions, disqualifying them from counting as genuine explanatory postulates or referring terms. Or so the dichotomy of the images suggests.

The dichotomy also suggests that the scientific image underlies the manifest image in that the latter can only exist in virtue of the existence of the former. Since the scientific image is often just a synonym for the image of the world sketched by the natural sciences, it follows that the perspective of the common man is ontologically inferior to, or derivative of, that which is offered by the natural sciences. Eddington's 'two tables' example helps illustrate this view. In his Gifford lectures in 1927 Eddington, apparently sitting behind a single table, presented his audience with two tables:

> One of them has been familiar to me from earliest years. It is a commonplace object of that environment which I call the world (...) if you are a plain common-sense man, not too much worried with scientific scruples, you will be confident that you understand the nature of an ordinary table (...) Table No. 2 is my scientific table. It is a more recent acquaintance and I do not feel so familiar with it. It does not belong to the world previously mentioned — that world which spontaneously appears around me when I open my eyes (...) I need not tell you that modern physics has by delicate test and remorseless logic assured me that my second scientific table is the only one which is really there — wherever 'there' may be. On the other hand I need not tell you that modern physics will never succeed in exorcising that first table — strange compound of external nature, mental imagery, and inherited prejudice — which lies visible to my eyes and tangible to my grasp. (Eddington 1928, pp. xi-xiv)

S. Clarke and T.D. Lyons (eds.), Recent Themes in the Philosophy of Science, 175–192.

My difficulty with the image of the images is that it seems to deny that science ultimately depends on common sense. It seems to imply that, although common sense gives rise to the practice of science and thus to hypotheses about the underlying structure of the phenomenological world, in the end the manifest image of common sense is radically false. Hence Eddington's remark that the scientific table is the only table which is really 'there'. The image of images can only do this, I will argue, if we take science to be something that has nothing to do with the initial stages of inquiry that (historically) gave rise to what we normally call science. In what follows, I will try to undermine the distinction between the manifest and scientific images, arguing from the pragmatist notion of concept elucidation, to show that the distinction between the images depends on a biased view of science. Using the resources of a pragmatist view on science I will argue that there is but one table in front of Eddington. This is indeed the scientific table, but it is not a table that is described exclusively by the natural sciences.

Perhaps I should add here my motivation for discussing the distinction and for taking the pragmatist turn. I am motivated to attend to the distinction in response to Jay Rosenberg's '*Nachruf* for Wilfrid Sellars' — Sellars died in 1989. Rosenberg states that:

> [a] leading challenge for contemporary philosophy consequently becomes to show how that tension [between the images] can properly be resolved, not by asserting the exclusivity of one image or the other but by a 'stereoscopic understanding' in which the two images come to be 'fused' into a single *synoptic vision* of man-in-the-world.
>
> (Rosenberg 1990b, p. 5)[1]

The reason I'd like to take the pragmatist turn has to do with the fact that both the authors whom I will discuss, with regard to the dichotomy of the images, Hilary Putnam and Jaap van Brakel, consider themselves to be inspired by the pragmatist tradition. More importantly, if one takes this tradition truly to heart, one will see the awkwardness of the image of the images.

Both Putnam and van Brakel are thinkers, driven by the work of the American pragmatists as well as Husserl and the later Wittgenstein, who have done a lot to debunk the myth of the images. In the next two brief sections I will discuss their attempt to either modify the idea of the priority of the scientific image (van Brakel 1996a) or solve the problematic relation between the scientific and manifest image by reintroducing a natural realism (Putnam 1994). I will argue that, in the case of van Brakel, shifting the priority from the scientific to the manifest image is ineffective in dismantling the dichotomy.[2] In the case of Putnam, I will suggest that natural realism is not enough to consign the dichotomy to oblivion.

Ultimately my suggestion would be to stop using the terms manifest and scientific image. It is too misleading to think that these terms connote with anything in the world. The word 'image' is far too metaphorical to be informative, and the terms 'manifest' and 'scientific' suggest a discontinuity between common sense and science that cannot be justified. I will argue that, if we reconsider the original, shared motivation of such diverse pragmatists as Peirce and Dewey (I will not consider James' work here), to take science as the final stage of inquiry, we will see that there is no need to think in terms of the images. Furthermore, there is no reason to think that there is a clash between 'common sense ontology' and 'scientific ontology.'

2. SELLARS: ORIGINAL, MANIFEST AND SCIENTIFIC IMAGE

We customarily refer to the work of Wilfrid Sellars when we speak about the manifest and scientific image. In fact, Sellars (1963) introduced a triad of terms, *original*, *manifest* and *scientific* image. I will not attempt to analyse Sellars classical contribution here[3], but I will point out that, according to Sellars, the manifest image is itself already a type of scientific image, as it is a refinement of the original image:

> [The manifest image] is not only disciplined and critical; it also makes use of those aspects of scientific method which might be lumped together under the heading 'correlational induction'. There is, however, one type of scientific reasoning which it, by stipulation, does *not* include, namely that which involves the postulation of imperceptible entities, and principles pertaining to them, to explain the behaviour of perceptible things.
>
> (Sellars 1963, p. 7)

Sellars admits that the proper name for the scientific image is the 'postulational' or 'theoretical' image (*ibid.*). He stresses that the distinction between the manifest and the scientific image is *not* that between 'an *unscientific* conception of man-in-the-world and a *scientific* one, but between a conception which limits itself to what correlational techniques can tell us about perceptible and introspectible events and that which postulates imperceptible objects and events for the purpose of explaining correlations among perceptibles' (p. 19). The contrast is therefore not between 'a pre-scientific, uncritical, naive conception of man-in-the-world, and a reflected, disciplined, critical — in short a scientific — conception' (p. 6). The contrast is rather that between a correlational scientific view and a theoretical-postulational scientific view.

Since my emphasis will not be on the issue of scientific realism-after-van-Fraassen,[4] I will ignore the Sellarsian notions of original, manifest and scientific images and, for the time being, go along with van Brakel (and Putnam) and take the manifest image to be the 'common-sense-human-life-form image'. However, if van Brakel states that the scientific image is concerned with neurons, DNA, and quarks, I protest and I protest by referring to the American pragmatist work of Peirce and Dewey. The scientific image, or rather 'science', is not concerned exclusively with neurons, DNA, and quarks.[5] Science acquires its character in relation to other methods of belief fixation, not in relation to its subject matter or content.[6] My modest aim is to call attention to this aspect of pragmatism and to help free ourselves of the obscure image of the images.

3. VAN BRAKEL'S QUESTION OF PRIORITY

Van Brakel says:

> The Manifest/Scientific Image terminology stems from Sellars (1963), but I broaden the meaning of 'manifest' intended by Sellars towards 'manifest social practices' and narrow it by keeping away from any association of 'manifest' with phenomenalism or sense-data. The manifest image is the daily practice or common-sense-human-life-form and includes references to such things as water, cats on mats, and being angry about an injustice; the Scientific Image is concerned about neurons, DNA, and quarks. Manifest forms of life are the manifest worlds of meaningful situations and behaviours in which we (already) live and which form the background for all scientific, philosophical, and all other activities.
>
> (van Brakel 1996a, pp. 259-60)

In his extensive article van Brakel addresses the split between the manifest, pre-theoretical image of the world and the scientific image of the world and the problem it evokes. This is the problem of how the ontology of the natural sciences underlies the ontology of the common man, the problem of how everything fits into one world. One way to solve this problem is to argue that the manifest image is simply wrong and that the terms used within this image do not refer to entities in the world. Usually, however, the idea of supervenience is introduced to account for this oneness of the world described in two radically different, but equally legitimate images.

The idea of supervenience is, roughly, the idea that physical facts determine all the facts: if the world were to be described differently in the manifest image, it would also have to be described differently in the scientific image. More importantly, supervenience is sometimes referred to when arguing for reductionism-without-elimination, when one is trying to preserve the manifest image, given the scientific image. I think that van Brakel (1996a) is right to describe the notion of supervenience as inadequate in this context, because it does not *explain* the relation between the manifest image and the scientific image. If two objects cannot differ with respect to their *A*-properties without also differing with respect to their *B*-properties, then properties of type *A* are supervenient on properties of type *B*. However, the relationship between *A*-properties and *B*-properties does not *explain* how, for example, the behaviour of elementary particles described in the scientific image underlies the behaviour of tables, as described in the manifest image. It just *states* that, *if* the property of being a table supervenes on the properties of elementary particles, then the behaviour of tables would be different if the behaviour of elementary particles were different.

In his evaluation of the many uses of the notion of supervenience, van Brakel concludes that 'there's a large discrepancy between what supervenience is expected to achieve and what the proposed definitions of supervenience actually offer' (1996a, p. 269). One would do well to rethink the relation between the manifest image and the scientific image, given the Sellarsian search for a synoptic vision, and van Brakel offers us two case studies to aid us in doing this. These are studies of schizophrenia and colour.

In the case of schizophrenia van Brakel argues that 'the reference of 'schizophrenic' is determined by *clinical* symptoms, which refer to the *manifest* characteristics of schizophrenia (...), *not* to brain or genetic properties' (p. 271), and in the case of colour he concludes that 'the only way to identify 'red' at the base level [the level of firing neurons] is to check what correlates with saying that something looks red' (p. 274). In both cases this means that 'everything worth talking about depends on the categories and intuitions embedded in the manifest image'. Van Brakel tells us that 'it is the manifest image that underlies the Scientific Image, not the other way around' (p. 278) and that 'the manifest image provides the foundations for the explanatory framework of science' (p. 289).

Besides the fact that van Brakel leaves the very distinction between the scientific and manifest image intact (something which is not to be expected if one takes a pragmatist attitude), his discussion leads to a rather trivial insight, namely that our investigations are *our* investigations. That is to say, we engage (already) in a world in which, for example, people act strangely. If we want to know why these people act strangely, we first try to mark and classify their behaviour, in order to determine our subject of inquiry. Next we try to find an explanation for their behaviour (schizophrenia). Finally, we discover that schizophrenic behaviour cannot be explained solely in terms of genetic or brain deficits and we modify our explanations.

To my mind, the mistake van Brakel makes in this sketch is to take the scientific image to be the Sellarsian *postulational* image presented by the sciences. Scientists postulate *prima facie* inaccessible, imperceptible, underlying structures. However the manifest image is the daily practice, or common-sense-human-life-form, where such postulational activity is simply lacking, or so says van Brakel. The result is a dichotomy of images and a priority dispute with no prospect of a fusion of images.[7]

But we can forego talking in terms of manifest and scientific images, forego talking in terms of the priority of the images and forego talking in terms of images at all. We can do this when we realise that science acquires its character in relation to other methods of belief fixation, and not in relation to its subject matter or content. The scientific 'image' is not opposed to the manifest 'image' in its postulational character (although it is a matter of fact that we do postulate underlying structures in our ongoing investigations of the world). Furthermore, science *as a specific method of belief fixation* does not invoke the idea of an image that clashes with a so-called manifest image — at least this is what a true pragmatist argues, and this is what I will argue for in Section 5. But before turning to classical pragmatism, let me first consider Putnam's point of view.

4. PUTNAM'S APPRAISAL OF SCIENCE

Even though Putnam's philosophical thought has witnessed major modifications, there seems to be at least one consistent strand in his thought. This is the idea that we should take science *seriously*, although not as an *absolute* description of the world.

So, for example, in his turn to pragmatism and its connection to the ideas of the later Wittgenstein, Putnam says:

> I do not, of course, wish to say that positrons aren't real. But believing that positrons are real has conceptual content only because we have a conceptual scheme — a very strange one, one which we don't fully 'understand', but a successful one nonetheless — which enables us to know what to say when about positrons, when we can picture them as objects we can spray and when we can't. (Putnam 1992, p. 60)

Even in his *Reason Truth and History* (1981), where he defended an internal realism, Putnam says that one can intelligibly claim that there are electrons, just as one can intelligibly claim that there are rabbits, since we use a criterion of reality that is relative to some conceptual scheme (1981, p. 52). Electrons are as real as rabbits. In 'Realism Without Absolutes' his claim is that, after taking the natural realistic or direct realistic turn, we can refer to a cat by its name 'because we can see the cat, and pet her, and many other things' (1993, p. 284). It seems that Putnam's recent plea for a *second naiveté*, in his *Dewey Lectures*, delivers us from scepticism and makes it possible to accept the scientific image of the world while *simultaneously* accepting 'the natural realism of the common man' (1994, p. 38). It does so by showing the '*needlessness* and the *unintelligibility* of a picture that imposes an interface [of sense data] between ourselves and the world' (1999, p. 41). Putnam revives the idea that to sense a table is 'to see that it is a (…) table that is in front of me' (1999, p. 14).

A *second naiveté* means (among other things) that for us to understand the word 'coffee table', is for us to use the phrase 'coffee table' by moving successfully within our language game. Moving around in the language game makes it possible for me to say that there is a coffee table in front of me. Putnam endorses the Wittgensteinian insight that 'understanding is having the abilities that one exercises when and in using language' (1999, p. 15).

But here one's reaction could be that there is a conflict between the natural realistic view and science. Science, the objection goes, tell us that there are no such things as coffee tables, only conglomerates of particles. Rebutting this claim is one of the aims of Putnam's *Dewey Lectures*. Putnam concludes by showing that neither the 'scientific image' nor the 'manifest image' can be absolutized. He argues that 'seeing that there is no conflict between natural realism and science, and no conflict between a suitably commonsensical realism and science' is crucial for steering us into the direction in which he wants us to go, viz. away from the idea that 'the form of all knowledge claims and the ways in which they are responsible to reality are fixed once and for all in advance' (1999, p. 20).

I can imagine how one could argue for a direct realism about unobservables. As with the case of the word 'coffee table', we can argue that we understand the word 'electron' by moving within our language game (for instance saying that electrons are too small to be seen). This allows me to say that there are really electrons. But the loose notion of a language game may hide the real issue at stake in direct realism namely the issuc of the status of, and relation between, the two 'images' somehow involved in that language game. Here is what Putnam has in mind:

> Mind talk is not talk about an immaterial part of us but rather a way of describing the exercise of certain abilities we possess, abilities that supervene upon the activities of our brains and upon all our various transactions with the environment but that do not have to be reductively explained using the vocabulary of physics and biology, or even the vocabulary of computer science. The metaphysical realignment I propose involves acquiescence in a plurality of conceptual resources, of different and not mutually reducible vocabularies (an acquiescence that is inevitable in practice, whatever our monist fantasies) coupled with a return not to dualism but to the 'natural realism of the common man'. (Putnam 1999, p. 38)

What this passage shows is that Putnam wants to retain the image of the images. To have the manifest image and its vocabulary and the scientific image and its vocabulary as mutually irreducible vocabularies, is 'inevitable in practice'. It looks like Putnam does not fully accept the pragmatist point of view regarding what it means to fixate our beliefs in a learning process of inquiry; what it means to accept the doubt/inquiry model of Peirce and Dewey. I will now return to Peirce's and Dewey's idea of inquiry, to answer the pressing question of the relation to the commonsensical view of the world — which Putnam seems to equate with the manifest image — and the scientific image of the world. Is the dichotomy intelligible, and if so, does one of the images have priority over the other?

My suggestion, in a nutshell, is to say, with Peirce and Dewey, that the great divide between the scientific image and the manifest image or 'the image of the common man' can be made to disappear. It can be replaced by the pragmatist notion of scientific *attitude*, resulting in a 'view', or hypothesis, in which the real furniture of the world is characterised by means of Peirce's criterion for what it means for something to be real. As is well known, Peirce has it that 'the opinion which is fated to be ultimately agreed to by all who investigate, is what we mean by the truth, and the object represented in this opinion is the real' (1878, p. 139). This may be interpreted as follows. All that is robust, in the sense that we need to refer to it for the sake of explaining or understanding the phenomena, is what makes up the ontology of the world and this is what we will find out if we take the scientific attitude. If we do not take this attitude we will possibly be left with mistaken (naive) beliefs, but not with a 'manifest image'.

The idea that there is a (prevailing) 'scientific image' in contrast to a (derivative) 'manifest image' is completely unclear, as is the opposing idea, that the 'manifest image' has priority over the 'scientific image'. According to pragmatists, we can and must adopt a scientific attitude and we will (perhaps) arrive at a view of our world that will contain electrons *and* coffee tables, as long as we need reference to both electrons and coffee tables, to make sense of the world we live in. My point with regard to Putnam's realism, briefly put, is that the commonsensical man must be (and cannot be anything less than) an *investigator*, or else the 'naive and natural realism of the common man' may turn out to be too naive. Inquiry implies giving up the idea that we can intelligibly use the terms 'scientific' and 'manifest' image and simultaneously say what is really there. For what is real is something we are in a constant process of finding out. What is real can only be decided tentatively.

In order to substantiate my claim, I will briefly discuss Peirce's and Dewey's pragmatism. I will focus on their shared anthropology in which they see man essentially as a community of inquisitors.[8] I will therefore neglect, *inter alia*, the important contribution due to Mounce (1997) who (like Rorty[9]) emphasizes the differences between Peirce's and Dewey's pragmatism. I will call attention to the pragmatic theory of meaning which, in combination with the pragmatic view of belief, results in the view that it is the fate of man to be an investigator. The naïve, natural realism of the common man, according to which there are coffee tables, in an important sense must and always will fit in with what is accepted and believed, after taking the scientific attitude.

5. PEIRCE AND DEWEY: MEN OF INQUIRY

Both van Brakel and Putnam accept pragmatism as the most viable philosophy, but they neglect (cf. van Brakel 1996b) or fail to appreciate in full (Putnam 1992, pp. 68-74) its notion of inquiry and belief. Isaac Levi claims that, despite the many differences between Peirce's, James' and Dewey's pragmatist view of man, they all share 'a preoccupation with the clarification of problem-solving inquiry based on a generalization of the principle of doxastic inertia to attitudes other than full belief' (Levi 1998, p. 181). Peirce's and Dewey's ideas about 'inquiry' and 'belief' suggest that, although their respective 'positions', realism and experimentalism, do not seem to share anything at all, there is a common ground on which they stand. It is to their notion of inquiry and belief I will now turn to, in order to illustrate how pragmatism rejects the idea that scientists have a scientific image of the world, whereas the common man has a manifest image of the world. I will summarize Peirce's late nineteenth century classics (1877, 1878) and discuss Dewey's turn of the twentieth century contribution on stages of logical thought (1900).

5.1. Peirce and the Fourth Fixator

In his 'Fixation of Belief' (1877) Peirce defends the claim that since 'irritation of doubt causes a struggle to attain a state of belief', the object of any inquiry is the settlement of opinion (p. 114). He then investigates several methods of belief fixation. It is important to understand that Peirce implies a particular anthropology here: we are such that we experience doubt as an irritation, and so we can do nothing but to try to get rid of this irritation and enter into a struggle to find a soothing state of belief. The sole object of inquiry, therefore, is 'the settlement of opinion' (p. 115) by attaining a state of belief in which we are 'entirely satisfied, whether the belief is true or false' (*ibid.*). Peirce then distinguishes his four methods of belief fixation: the method of tenacity, authority, the a priori method and the scientific method.

Every method distinguished has its advantage and its disadvantage (see Table 1). Although Peirce does not claim we should all abandon the method of tenacity, of authority or the a priori method in favour of the scientific method (p. 121), he does suggest that the method of science is like a 'chosen bride' and that we must 'strive to be the worthy knights and champions of her' (p. 123) since 'scientific investigation has had the most wonderful triumphs in the way of settling opinion' (p. 120).

METHOD	PRO	CON
TENACITY	Evades irritation of doubt efficiently	Social impulse is against it
AUTHORITY	Fixes belief in the community	Some men possess a feeling of contingency about their beliefs
A PRIORI	Respectable from the view of reason	Makes inquiry something similar to the development of taste
SCIENCE	Ultimate conclusion of every man will be the same	

Table 1: Peirce's Methods of Belief Fixation[10]

The driving force behind the shift from one method of belief fixation to another method of belief fixation is the recognition of the element of contingency in our beliefs. The recognition of the fact that in the *a priori* method our ideas depend on something contingent, namely the contingency of the chosen style of reasoning, brings us to the idea that our beliefs must to the contrary be based upon an 'external permanency' (p. 120). Postulating such an external permanency is what characterizes the scientific method.

The method of science is to make our beliefs independent of our thoughts. In fact it comprises a hypothesis: 'There are real things, whose characters are entirely independent of our opinions about them; those realities affect our senses according to regular laws, and, though our sensations are as different as our relations to the objects, yet, by taking advantage of the laws of perception, we can ascertain by reasoning how things really are, and any man, if he have sufficient experience and reason enough about it will be led to the one true conclusion' (p. 120).

In the follow-up paper 'How to Make Our Ideas Clear' (1878) it becomes clear that 'the opinion which is fated to be ultimately agreed to by all who investigate, is what we mean by the truth, and the object represented in this opinion is real' (p. 139), so that we are led to believe that the scientific method will fixate, in the long run, on true beliefs. But we detect a very peculiar characterization of belief in Peircean pragmatism. 'And what, then, is belief?', Peirce asks his reader. Most importantly belief 'involves the establishment in our nature of a rule of action, or, say for short, a *habit*.' (1878, p. 129)

This belief about belief implies, in an intuitive way, Peirce's pragmatic maxim. Since I strive for beliefs that will survive future situations into which I will be taken, I will also strive for a certain degree of clarity of the concepts I use in as much as circumstances call for clarification. Peirce's pragmatic maxim (his theory of concept elucidation) can be formulated as follows: if you want to know what our idea of anything is look for our idea of its sensible effects (p. 132). This maxim seems to flow naturally from his idea about ideas. Let me illustrate the maxim at work with an example.

Suppose I introduce you to a certain woman in my vicinity by saying 'I'd like you to meet my mother'. Calling her my mother is, to me, a habit. Believing that she is my mother is a habit and it makes me introduce you to her as a rule of action. I do not have any motivation to consider the meaning of 'mother', until I am confronted with a problem which stirs in me the sensation of doubt when she says 'Son, you are not my child'. Now it becomes important to me to know what we mean by 'mother'. If I want to take away my doubt about whether she is or is not my mother, I consider one or more of the methods of belief fixation. If I aim for true beliefs concerning the concept of mother I must use the scientific method and efficiently explore the 'practical bearings' of the concept of mother, given that there is 'an external permanency'.

I will not ignore her statement (method of tenacity). I will not refer to my birth certificate (method of authority). I will not arbitrarily say that 'mother' is a pedagogical (and not a genetic) concept and therefore believe that this woman is my mother (a priori method). I will, rather, postulate an external permanency and conclude that I must accept that there is simply no relevant genetic bond between me and her, as the natural sciences teach me. I am not born out of her womb, but I know that she raised me, and the human sciences teach me that the social bond between us has determined my being in the world. This social bond is a reference to an external permanency. For all humans who have such a social bond, the effects are more or less the same. The relation between this woman and me is posited as real and of a specific kind. We cannot (now) understand my behaviour unless we consider this social relation as referring to an external permanency, to something that is independent from thought. This analysis is conjectural of course. Ultimately the practical bearings of the concept of mother will make it clear what our idea of mother is — genetic engineering, cloning, artificial intelligence, different forms of community life: they may all have consequences for our idea of the concept of mother — and whether we need the idea of mother to explain our behaviour. For now this analysis may suffice to settle my doubt that this woman is or is not my mother.

Does this scientific attitude, pragmatically interpreted, result in a dichotomy of images? It does not. Must we now say that my mother appears in two opposite images? Does my mother appear in the scientific image as the woman to whom I have no genetic but only a specific social relationship, and in the manifest image as 'my mother'? Clearly not. Are we now confronted with two images that need to be fused in order to attain a Sellarsian synoptic view? No. The situation is of a completely different nature due to the doubt/belief notion of inquiry. I conclude that the concept of mother (as far as we now know) must appear in the scientific *hypothesis* or *view* of our world, and not in either the 'manifest image' or the 'scientific image'. We are in a constant process of finding out what our idea of mother is, by finding out what the sensible effects of the concept are.

The same holds for Eddington's table: our idea of table is our idea of its sensible effects, and that means that the table does not appear in the 'manifest image' as the 'ordinary' table and in the 'scientific image' as the 'physical' table, but rather that the concept of table appears in the scientific conjectural hypothesis of our world. We are always in a process of finding out what our idea of table is, and we think about it harder when we are confronted with a doubt concerning tables. We use the scientific method in fixing our beliefs. We do not use the scientific method to create a scientific image, in contrast to a manifest image, and nor is the result of scientific inquiry such an image of images — or so the true pragmatists argue.

A lot more could be said about Peirce's two classical papers, but I will conclude here by sketching a pragmatic anthropology. Peirce draws the following picture. What people believe is expressed by what they do or say, and what they would do or say in various circumstances. Whether what they believe is true or false is ultimately (in Peirce's 'long run') determined by using the scientific method, which postulates a reality (an external permanency). It is by this method of belief fixation that we learn *par excellence* by discovering the practical bearings of our concepts. We test our concepts and beliefs about the world in a community of inquisitors, and that reality appears to us to be independent of the individual investigator (but not independent of the community of investigators). True knowledge is the 'opinion which is fated to be ultimately agreed to by all who investigate' and real is what 'is represented in this opinion' (1878, p. 139).

According to Peirce, this is the way reality and truth must be interpreted if we want to break away from a dogmatic Cartesian epistemology. Humans are fallible creatures and of such a nature that they evidently strive for 'thought at rest'. Pragmatism thus suggests an anthropology in which humans appear as finite, yet able to learn from mistakes. This anthropology may be connected to the history of America (Menand 2001) or traced back to the rise of scientific knowledge itself. However, for Peirce, pragmatism is deduced from the idea that it seems to be the only way to make sense of the world of phenomena.

Due to the pragmatist rendering of inquiry, belief and clearness of ideas (the pragmatic maxim) to even pose the question whether, for instance, 'mothers can be reduced to the behaviour of conglomerates of particles' means first clearing up all the obscure meanings involved. If we do that, Peirce would hold, we will discover over time what we need to postulate (and understand as real) in order to understand phenomena. Similar reasoning holds for Dewey (1900).

5.2. Dewey and the Aim of Science

John Dewey agrees with Peirce that the purpose of thinking is to secure a stable equilibrium and he explicitly opts for the Peircean doubt/inquiry model. In his 'Supremacy of Method', Dewey expresses the opinion that, although many definitions of mind and thinking have been given, there is only one 'that goes to the heart of the matter: — response to the doubtful as such' (Dewey 1929, p. 345). Dewey strives to show 'some of the main stages through which thinking, understood in this way, actually passes in its attempt to reach its most effective working; that is the maximum of reasonable certainty' (Dewey 1900, p. 183). These stages of belief fixation nearly correspond on a one-to-one basis to Peirce's four stages. Dewey's work illustrates how this concept of inquiry implies an imageless view of man-in-the-world.

Beliefs bring doubt to a halt. At first, we take ideas or beliefs as fixed. Ideas have a static and rigid meaning and their function is to solve conflicts. They represent 'a recognition of a habitual way of belief: a habit of understanding' (1900, p. 187). An idea, therefore, is 'a scheme of assigning values', or 'a way of dealing with doubtful cases'. This first stage of seemingly fixed ideas may be called *dogmatic*. When the whole of our fixed ideas and beliefs become more complex, however, we feel a need for discrimination of, and reflection on, our ideas. This brings us to a second stage of belief fixation: the *critical stage*.

Complexity presents us with 'critical cases', cases we cannot solve by using rigid and fixed ideas. We ultimately conclude that the fixation of our beliefs is not inherent to these beliefs but the result of *our* workings. The result is discussion and dialogue, a conversation of thoughts. We then realise that we must find a method of belief fixation that offers us a new effective way of handling problematic situations, as discussion only leads to a 'clash of ideas', and not stability. This instability is historically followed by the *axiomatic stage*.

Dewey describes this transition as the transition of discussion into reasoning, of 'subjective reflection into method of proof' (p. 199). This method of proof is modelled on the syllogism. If we accept the premisses and if we accept the way of reasoning as a valid one, this method restores the stability of our beliefs. But it also leads, ultimately, to a restless feeling of relativism, to a feeling that it is only taste that determines our beliefs. Our beliefs are only conditionally fixed. Although 'doubt is awake and inquiry is active', this inquiry is actually 'rigidly limited' (p. 206).

The transition to the fourth method of belief fixation gives us *science*. We witness the transition of the method of proof to the method of inference. This method of inference is a procedure of discovery, a procedure that takes us from the known to the unknown. It is a method 'for making friends with facts and ideas hitherto alien' (p. 210). Science affords us to direct our inquiry in every direction *we may wish for*. In that sense science is discovery rather than justification (p. 208). Within science, suggestions are not so much assessed by the degree to which they fit in with our accepted theories, but to the degree in which they can lead to the discovery of new knowledge about the world. Science, for Dewey, is 'inquiry emancipated (...) whose sole aim and criterion is discovery' (p. 216). Science eases our thoughts and soothes our minds since we realize that science is done for the very fact of initiating inquiry, which is our way to make our ideas clear and helps us cope with the hazardous world. This method is the natural outcome of a process of belief fixation if we link thought to doubt and inquiry the way Peirce and Dewey do. Dewey thus accepts the Peircean view on science (see Table 2).

METHOD	PRO	CON
DOGMATIC	Short cut to certainty	Complexity makes it unworkable
CRITICAL	Deals with complexity	Instability
AXIOMATIC	Restores stability	Beliefs are only conditionally fixed
SCIENTIFIC	Makes the unknown known	

Table 2: Dewey's Methods of Belief Fixation.

Peirce's and Dewey's shared concept of inquiry does not imply a dichotomy of images. Their notion of inquiry is concerned with a method of belief fixation and an attitude that we may call scientific. It is a happy surprise to see that Dewey, in his 'Supremacy of Method' uses Eddington's two tables example to discuss 'the problem which is supposed to exist between two tables, one that of direct perception and use and the other that of physics' (1929, p. 355). He argues that this problem is illusory:

> The perceived and used table is the only table, for it alone has both individuality of form — without which nothing can exist or be perceived, and also includes within itself a continuum of relations or interactions brought to a focus. (Dewey 1929, p. 355)

What Dewey calls *the* table 'as *not* a table but as a swarm of molecules in motions of specified velocities and accelerations', that is, the table as a physical object, the table as it appears in the natural sciences, is an abstraction which 'unless it is hypostatized' is not a vicious abstraction (p. 354). *The* object designates 'selected relations of things which, with respect to their mode of operation, are constant within the limits practically important' (p. 354). And again, this pragmatic stance does not imply two images, nor two ontologies. It relates science to a method of settling doubt, rather than a preconceived idea about the content of science, and thus it seems to circumvent the problem of fusing the images into a synoptic view of man-in-the-world.

6. CONCLUSION: WITHOUT IMAGES

I think we can safely conclude that, according to both Peirce and Dewey, the scientific method of belief fixation evades the problem of the two seemingly incompatible images. If we accept the classical pragmatist ideas about methods of belief fixation and use the doubt/inquiry model as a model applicable to science, then the methodical character of science comes into focus. On this view science does not invoke an insurmountable dichotomy of images. *Thus there is no scientific image in contrast to a manifest image.* There is only a *scientific attitude*, a scientific method of belief fixation, which leads to a hypothesis in which our so-called common sense notions will appear if they refer to things robust enough to survive an endlessly selective process. The problem of how the idea of conglomerates of particles relates to the idea of, say, mother or table (the problem of supervenience) is something we will need to further clear up using the pragmatic maxim. My hope is that this issue can indeed be cleared up along pragmatist lines, but I fear that I must leave this as a promissory note.[11]

Tilburg University

REFERENCES

Aune, B.: 1990, 'Sellars's Two Images of the World', *Journal of Philosophy*, 87, 10, 537-545.

De Regt, H.C.D.G.: 1996, 'The Second Best Approach to the Problem of Scientific Realism: the Rationality of Belief', in L. Horsten & I. Douven (eds.) *Realism in the Sciences*, Louvain Philosophical Studies 10, 87-113.

De Regt, H.C.D.G.: 1999, 'Pragmatism, Scientific Realism and the Problem of Underdetermination', *Transactions of the Charles S. Peirce Society*, Spring, Vol. XXXV, No. 2, 374-397.

De Regt, H.C.D.G.: 2001, 'The Hybrid Science of Psychology. A Minimal Contribution from Philosophy of Science', in A.A. Derksen (ed.) *Moving Ahead: Philosophy of Mind and Realism*, Dutch University Press, Oisterwijk, pp. 105-119.

De Regt, H.C.D.G.: 2002 'Restless Thought: A Pragmatist View on Belief and Scientific Realism', in G. Debrock (ed.) *The Quiet Revolution. Essays in Process Pragmatism*, Rodopi (to be published).

Dewey, J.: 1900, 'Some Stages of Logical Thought', in J. Dewey, 1916, *Essays in Experimental Logic*, Dover Publications, New York, pp. 183-219.

Dewey, J.: 1929, 'The Supremacy of Method', in his *Quest for Certainty*, Minton, Balch & Co., New York. Cited from Fisch 1996, pp. 344-360.

Eddington, A.S.: 1928, *The Nature of the Physical World. Gifford Lectures 1927*, Cambridge University Press, Cambridge.

Fisch. M. (ed.): 1996, *Classic American Philosophers*. Fordham University Press, New York.

Gouinlock, J.: 1990, 'What is the Legacy of Instrumentalism? Rorty's Interpretation of Dewey', *Journal of the History of Philosophy* XXVIII, 2, 251-269.

Lange, M.: 2000, 'Salience, Supervenience, and Layer Cakes in Sellars's Scientific Realism, McDowell's Moral Realism, and the Philosophy of Mind', *Philosophical Studies*, 101, 213-251.

Levi, I.: 1998, 'Pragmatism and Change of View', *Canadian Journal of Philosophy Supplementary Volume* 24, pp. 177-202

Menand, L.: 2001, *The Metaphysical Club. A Story of Ideas in America*, Farrar, Straus, and Giroux.

Mounce, H.O.: 1997, *The Two Pragmatisms. From Peirce to Rorty*. Routledge, London.

Peirce, C.S.: 1877, 'The Fixation of Belief', in: N. Houser & CH. Kloesel (eds.) *The Essential Peirce. Selected Philosophical Writings. Volume 1, 1867-1893*, Indiana University Press, Bloomington, pp. 109-123.

Peirce, C.S.: 1878, 'How to Make Our Ideas Clear', in: N. Houser & Ch. Kloesel (eds.) *The Essential Peirce. Selected Philosophical Writings. Volume 1, 1867-1893*, Indiana University Press, pp. 124-141.

Putnam, H.: 1981, *Reason, Truth, and History*. Cambridge University Press, Cambridge.

Putnam, H.: 1992, *Pragmatism*. An Open Question. Blackwell, Oxford.

Putnam, H.: 1993, 'Realism Without Absolutes', cited from Putnam 1995.

Putnam, H.: 1994, 'Sense, Nonsense, and the Senses: An Inquiry into the Powers of the Human Mind' (The Dewey Lectures 1994), *Journal of Philosophy*, vol.91, nr.9; cited from Putnam 1999.

Putnam, H.: 1995, *Words and Life*, Harvard University Press, Cambridge, Mass.

Putnam, H.: 1999, *The Threefold Cord. Mind, Body, and World*, Columbia University Press, New York.

Rorty, R.: 1980, 'Pragmatism, Relativism, and Irrationalism', *Proceedings of the American Philosophical Association*, LIII, 719-738.

Rosenberg, J.F.: 1990a, 'Response to Aune', *Journal of Philosophy*, 87, 10, 546-547.

Rosenberg, J.F.: 1990b, 'Fusing the Images. Nachruf for Wilfrid Sellars', *Journal for General Philosophy of Science*, 21, 1-23.

Sellars, W.: 1963, *Science, Perception and Reality*, Ridgeview Publishing Company, Atascadero, California.

Sellars, W.: 1976, 'Is Scientific Realism Tenable?', *PSA* 1976, vol. 2, 307-334.

Van Brakel, J.: 1996a, 'Interdiscourse or Supervenience Relations: The Primacy of the Manifest Image', *Synthese* 106, 253-297.

Van Brakel, J.: 1996b, 'Empiricism and the Manifest Image', in L. Horsten & I. Douven (eds.) *Realism in the Sciences*, Louvain Philosophical Studies 10, 147-164.

Van Fraassen, B.C.: 1976, 'On the Radical Incompleteness of the Manifest Image', *PSA* 1976, vol. 2, 335-343.

Van Fraassen, B.C.: 1994, 'The World of Empiricism', in J. Hilgevoord (ed.) *Physics and Our View of the World*, Cambridge University Press, Cambridge, pp. 114-134.

NOTES

[1]See also Aune (1990) and Rosenberg (1990a) on Sellars's distinction of the images, and Lange (2000) for a different and substantial attempt to contribute to a 'synoptic vision of man-in-the-world'.

[2]Van Brakel, to be sure, is sometimes more cautious when arguing that '*if* we want to raise the question of priority, it is the manifest image that controls the Scientific Image and not vice versa' (1996a p. 259).

[3]Cf. the PSA debate between Sellars (1976) and van Fraassen (1976).

[4]See my (1996) for a discussion on van Fraassen's *Scientific Image*.

[5]Of course, van Brakel is right if we mean by the scientific image the *Sellarsian* scientific image since Sellars introduced the image by *stipulating* that it includes *imperceptible* entities. My point here is that this 'scientific' image is not a product of scientific inquiry but of philosophers thinking (wrongly) about science.

[6]In his (1994) van Fraassen argues that if one accepts this 'then realism has throughout mis-focussed the debate [on scientific realism]' (p. 132). My rather blunt counter-argument would be that if van Fraassen accepts the 'primacy of method *vis à vis* content in science' then empiricism may have mis-focussed the debate *also*. But this obviously calls for further elaboration, which will have to be postponed.

[7]The question of how social and physical or biological levels of psychological discourse are to be unified is a problem that needs to be addressed, but it is a problem to be addressed in our ongoing scientific investigations. See my (2001) for a brief remark on the status of psychology as a hybrid science.

[8]Even when Putnam in his *Pragmatism* brings out the differences between the Carnapian and pragmatist picture of inquiry (1992, p. 70), he fails to see that the pragmatist notion of inquiry does not lead to the image of images, or the image of mutually irreducible vocabularies, but rather to the adoption of a scientific *attitude*.

[9]But see the devastating critique by Gouinlock (1990) of Rorty's interpretation (for instance in his 1980) of Dewey's work.

[10]In my (1999) I discuss a possible con of the scientific method in relation to the problem of underdetermination.

[11]I would like to thank my colleagues in the Dutch Tilburg/Nijmegen research programme Rationality and Non-Reductionism (especially Arno Wouters), Wayne Christensen, and the editors for their very helpful comments on an earlier draft.

BRIAN ELLIS

HUMAN AGENCY, REALISM AND THE NEW ESSENTIALISM

1. INTRODUCTION

The new essentialism[1] is a metaphysic for a species of scientific realism that embraces the theoretical entities of most of the accepted causal process theories of science, as well as the causal powers, capacities and propensities of the things that are necessarily involved in these processes. Let us call those who are realists about all such entities 'causal process realists'. It has been argued elsewhere (Ellis 2001, and forthcoming) that the new essentialism provides a sound basis for such a realist ontology. It entails that there are hierarchies of facts about the world that exist independently of anyone's knowledge or understanding. It explains the modal structure of the world, with its natural necessities, and its hierarchical system of laws of nature. It may not be necessary to accept the whole metaphysical stance of the new essentialism to be a causal process realist. For the ontology would appear to be defensible, even if one does not accept all of these metaphysical explanations. But, as Stathis Psillos (1999) has argued convincingly, a scientific realist must at least accept that the world has a natural kinds structure, and this is already a first step in the direction of essentialism.

The new essentialism not only updates the scientific image of the world to include hierarchies of natural kinds, it also changes the scientific image from that of an essentially passive world, to a dynamic one in which things have active causal powers, and are constantly interacting. In thus updating the scientific image of the world, the new essentialism provides a sound basis for an argument for causal process realism. This new argument is a form of the 'No Miracles Argument'. However, it proceeds from a different premiss from Putnam's famous argument, and it arrives at a somewhat different conclusion. The required premiss is that the new scientific image of reality is basically an epistemically right[2], and more or less adequate, de-

S. Clarke and T.D. Lyons (eds.), Recent Themes in the Philosophy of Science, 193–207.

scription of reality. That is, most of the kinds of things and processes that scientists believe in can properly be regarded as permanent features of the new scientific image, and there are no major components of the manifest image that it fails to account for which would require us to postulate the existence of other kinds of things or processes.

The question then is how best to account for the basic epistemic rightness of this new image. It will be argued here that there is really only one plausible explanation, viz. that the world really has a natural kinds structure of the sort described in the new image – so that the image is basically true of the world in the sense that it roughly corresponds to it. The argument to this conclusion will be elaborated in Section 4 below.

Much of this paper will be concerned with elaborating the new scientific image of the world, and arguing for its superiority over the passive one. Given the image of a passive reality, there is a plausible objection to realism concerning this image. But while this objection is a powerful one against the old scientific image of reality, it is powerless against the new one. The new scientific image of the world makes clear and obvious sense of the phenomena of human agency.

2. THE NEW SCIENTIFIC IMAGE

Wilfrid Sellars (1963) speaks of two very different images of reality. He calls them the 'manifest' and the 'scientific' images. The manifest image is the view of ourselves, the world, and of our place in it, which derives from common experience, and from critical reflection on that experience. The scientific image is the picture of the world that science yields. The two images are not obviously compatible. The scientific image, as it is presented to us by scientific realists, is objective, but seemingly dead and impoverished; the other, the manifest image, is often held to be dependent somehow on us as observers (or thinkers, or language-users), but it is a rich image, inhabited by living creatures, and things with genuine causal powers.

The scientific and the manifest images of reality should complement each other. If the scientific image is ultimately to be preferred, as I believe it should be, then the failings of the manifest image should be theoretically explicable. For the manifest image, being compiled more or less directly from the data provided by observation and introspection, is a crucial test of the adequacy of our scientific image of the world. If we are being systematically misled by observation or introspection, then we need to know how and why, because, in the absence of any such explanation, there would be no sufficient reason to take the scientific image of the world seriously.

Individual theories that are successful in their own areas might reasonably be supposed to be true or approximately true, as most scientific realists argue. But if the overall picture is irreconcilable with the way the world appears to us to be, then the scientific enterprise has manifestly failed to account adequately for some of the data of experience, and something is seriously wrong somewhere.

In this paper, I want to elaborate a scientific image of reality that is not open to the serious charge of being inadequate to account for human agency. Specifically, I wish to argue that our capacities for conscious decision-making, willing and acting, which are often supposed by philosophers to be incompatible with the view of mankind that science requires us to take, can nevertheless be reconciled with a scientific image of mankind. But the scientific image that I wish to defend is not the one that is usually presented to us by philosophers. For the philosophic image of the scientific image of the world is an impoverished one that few scientists nowadays would, or should, take seriously. It is an image that was originally inspired by seventeenth century mechanism, and reinforced by Hume, among others, in the eighteenth century. Although it contains some concessions to modernity, the present philosophic image of the scientific image is still basically eighteenth century in outlook.

The original scientific image of seventeenth century mechanism was of a passive world consisting entirely of atoms which, in themselves, were neither coloured nor colourless, hot nor cold, sweet nor bitter, nor any other perceptible quality. The atoms, of which all things were thought to be composed were supposed to be infinitely hard and impenetrable, but, apart from their impenetrability, they were thought to have only the mechanical properties of shape, size, and (sometimes) mass, and the capacity to move or change orientation when pushed. According to E.A. Burtt, the mechanistic world was one that was 'hard, cold, colourless, silent, and dead; a world of quantity, a world of mathematically computable motions in mechanical regularity' (1932, p. 237). The manifest image, by contrast, was not passive, but inhabited by things having real causal powers, and by living, thinking, experiencing, conscious beings whose actions were often intentional, and intended to shape the world around them to suit their purposes. The world of our experience could not, it seemed, be reduced to events occurring in the dead mechanistic world of science.

The most common response of seventeenth and eighteenth century philosophers to this problem was to divide the world into mental and physical components. Mental events were thought to be essentially different from physical events, to occur in different substances, and occupy different realms. Science was taken as providing a description of the physical world, but not of the world of our experience. We are ourselves not even in the scientific picture, they thought. The scientific account of reality would include descriptions of our bodies, perhaps, but it did not, and could not, also include descriptions of our inner selves or our experiences.

Such dualism did not solve the problem, however. If the material universe consisted of one kind of substance (having the primary characteristics of matter in a mechanistic world), but the human mind was made of a different kind of substance (having the capacities for experiencing, thinking, deciding, willing, and so on) then what is the relationship between the two? How can physical events produce mental events (e.g. in percept-ion), or mental events produce physical ones (e.g. in acts of will)? In which domain do the answers to these questions lie? Dualism thus created at least as many problems as it solved. It removed the need to provide a mechanistic theory of mind, but it provided no clue as to what an alternative theory would be like, or how its mental processes would be related to the mechanisms of the body.

Dualism may not be acceptable, but the scientific image, as it is usually presented to us by scientific realists today, is also unacceptable. For it presents what is still, essentially, a Humean view of causation, has no natural place within it for many of those most human qualities and capacities which inform the manifest image we have of ourselves as rational agents observing and responding to each other, and to the world around us. So, the big question is: How can these two very different images of reality be reconciled?

Of all of the problems of philosophy, this is perhaps the most intractable. It cannot be solved just by focussing on the manifest image, and attempting to articulate it. Nor can it be solved by resolutely attending to the nature of scientific inquiry, and ignoring its relationship to ordinary human experience. The two images must somehow be brought together, so that each can be seen in relationship to the other for what it is. The scientific image is far too powerful to be dismissed as a fabrication, with no implications for our conception of ourselves. On the other hand, the philosophers' picture of the scientific image, which has dominated Anglo-American philosophy since the eighteenth century, seems too bare and passive to yield a satisfactory account, even of causation. And it is manifestly inadequate to provide a sound basis for understanding human agency.

It will be argued here that the scientific image, as it has traditionally been portrayed by philosophers, is much more impoverished than it needs to be. For it represents an outdated view of reality. It does this by portraying inanimate nature as intrinsically passive, and therefore as being *prima facie* incapable of acting, except under the influence of external forces.[3] To bring the two images closer together, the scientific image needs to be improved. Specifically, it has to be recognised that the natural world is not intrinsically passive, but essentially active and interactive.

The new essentialism is a metaphysic in which this fact about nature is recognised as being fundamental. It is a metaphysic which promises to create a scientific image of the world that is very different from the Humean one which mostly dominates the thinking of scientific realists. The new essentialists' image is of an active world in which things have intrinsic causal powers; it is not that of a passive world of the kind in which Hume, his mechanistic predecessors, and his many followers, believed (and still believe). In an active world of the kind envisaged by scientific essentialists, all things have causal powers, and are therefore agents of one kind or another. So the power of agency is not something unique to human beings, or even to living creatures. It is a pervasive feature of reality. This is not to say that human agency is not something rather special: it clearly is. On the other hand, it is not as alien to the essentialists' view of the world as it is to the Humean one.

3. HUMEAN AND ESSENTIALIST PERSPECTIVES ON REALITY

Most philosophers today still believe, as Hume did, that the question of what causes what ultimately depends on what universal regularities hold. Their theory of causation thus makes it very difficult for them to account for human agency. If they are right, then our conscious decision-making processes, and the actions that we say stem from them, must all be understood in terms of regularities, constant conjunctions, and the like, concerning which we, as conscious beings, can be nothing other than introspective spectators. But this is clearly not how they are in the manifest image. We do not see ourselves as being in such a passive role. Rather, we see ourselves as acting and doing things for reasons. We see our processes of deliberation as ones that are thoroughly under our control, and which we can continue, suspend, or eventually act upon. Acceptance of a Humean theory of causation thus makes it very difficult for anyone even to suggest a plausible theory of human agency.

The new essentialism changes the scientific picture. For an essentialist, all effects are displays of causal powers, or due indirectly to such displays, (as is the darkening of the room when the blinds are pulled). And these effects are not just events that happen to follow the triggering of causal powers; they are their manifestations, or at least the consequences of their manifestations. If the mousetrap is not set off by the taking of the cheese, then presumably the disturbance was not great enough to release the causal power latent in the spring. Unless there are extraordinary defeating circumstances, there can be no question of the catch being released and the mousetrap not snapping shut. Such an unlikely event could only occur if something were to intervene to prevent the mousetrap snapping shut. In the absence of any such defeaters, the mouse will be a dead mouse.

Essentialists take the view that things have intrinsic causal powers that determine how they must act and interact with each other. The world is thus conceived to be a causally active place, in which things are naturally disposed to act and interact with each other in various ways, depending on how they are related to each other, and on what intrinsic causal powers they have. There are no superimposed laws of nature in the essentialists' view of reality. On the contrary, the laws of nature are considered to be immanent in the world, and to be determined by the essential natures of things.

According to the new essentialism, the world is structured into hierarchies of natural kinds of objects, properties and processes. It is not an amorphous world on which we must somehow impose our own categories. There is a pre-existing grid of objective categories, they argue, and it is the aim of natural science to reveal and describe them. The distinctions between the chemical elements, for example, are thought to be real and absolute. There is no continuum of elementary chemical variety, they say, which we must arbitrarily divide somehow into the chemical elements. The distinctions between the elements are there for us to discover, and the sharp distinctions between them are guaranteed by the limited variety of quantum mechanically possible atomic nuclei. Many of the distinctions between kinds of physical and chemical processes are also real and absolute. There is no continuum of processes within which the process of ß-emission occurs, and from which it must be arbitrarily distinguished. According to the new essentialism, the world is just not like that. At a fundamental level, the processes that occur often allow real and absolute distinctions of kind to be made. Therefore, if there are natural kinds of objects or substances, there are also natural kinds of events and processes. The new essentialism is thus concerned with natural kinds that range over events or processes as well as with those of the more traditional sort which range over objects or substances.

The natural kinds of these two types evidently occur in hierarchies. At the apex of the hierarchies, there are, it may be supposed, two very general natural kinds. The most general of the natural kinds in the category of objects or substances presumably includes every other natural kind of object or substance that exists, or can exist, in our world. This is the global kind, for our world, in this category. The most general kind in the category of events is the global kind that includes every other natural kind of event or process, which occurs, or can occur, in our world.

The laws of nature then relate to the kinds. The most general laws of nature, the global laws, describe the essential properties of the global kinds, and therefore hold necessarily of all members of these kinds. The law of causality, for example, which states that every object that exists, or could exist, in our world has intrinsic causal powers, is a global law of nature relating to the first of these global kinds. Another such law is the principle of physicalism, which states that every object which exists, or can exist, in our world is a physical object, i.e. an entity that has mass or energy. The law of conservation of energy, which states that every event or process is intrinsically conservative of energy, is a comparably general law relating to the global kind that consists of all physically possible events and processes.

This law implies that no event or process that is not intrinsically conservative of energy is metaphysically possible in our world. The laws we think of as causal laws are generally more specific in their direct application. The laws of electromagnetism, for example, apply to all electromagnetic fields, and hold necessarily of all such fields. But they do not apply to other kinds of field, and any field that is structured according to these laws must be electromagnetic. The laws of chemical combination are more specific still. These are the causal laws, par excellence.

The tenet of essentialism that perhaps sets it most clearly apart from Humeanism, is the principle of causality, i.e. the claim that everything that does or could exist in this world has intrinsic causal powers. For it implies immediately that the essential properties of the most fundamental kinds of things are not just the passive primary qualities of classical mechanism, but also include various causal powers – i.e. powers to act and interact. In other words, the basic things in the world are essentially active and dynamic. They are not just passive objects being pushed or pulled around by God, or by the contingent forces of nature, but things whose essential natures are determinative of their behaviour. The claim that the world is structured into hierarchies of natural kinds of objects, and so on, could, in principle, be accepted by Humeans. Things of different natural kinds, they might say, are just things made up of different basic ingredients, or of the same ingredients put together different ways. But there is nothing in their natures, they would add, which requires that they should behave in one way rather than another. How they are disposed to behave, they would say, depends on what the laws of nature happen to be.

Essentialists reject this claim. According to the new essentialism, all things are essentially active and reactive. At the most basic level, what they are intrinsically disposed to do is what makes them things of the kinds they are. Their identities as members of these kinds depend on their being so disposed to act. If this thesis of the new essentialism is correct, then the laws of nature are not contingent, as nearly everyone else supposes, but metaphysically necessary, and hence true in all possible worlds. That is, it must be *metaphysically impossible* for things, constituted as they are, to behave other than in accordance with the laws of nature. Even God (assuming Him to exist and to be all powerful) couldn't make them behave contrary to their natures. He might *change* their natures, perhaps, so that they might become, or be replaced by, things of different kinds. But there is no possible world in which things, *constituted as they are*, could behave any differently. For them to behave differently, they would have to be or become things of different kinds, or be made up of things of different kinds.

4. THE ESSENTIALIST ARGUMENT FOR REALISM

The new essentialist's argument for realism about the theoretical entities of science, and their causal powers, capacities and propensities, is an argument to the effect that there is no rational alternative to a broad endorsement of realism about such entities. It is not the same as Putnam's argument, however, because it has a different explanandum, and a different explanans. Only the gross form of the argument is the same.

1. According to the essentialist perspective on scientific understanding:

   (a) The world is an elaborate, strongly interconnected, hierarchical structure of distinct kinds of objects (particles, fields, etc.).

   (b) These objects have various causal powers, capacities and propensities, which drive the natural processes in which they are involved.

   (c) The laws of nature describe these objects and their properties, and have the same sort of hierarchical structure as the kinds of entities over which they range.

The essentialist perspective on scientific understanding thus yields the new scientific image of the world.

2. The new scientific image of the world has only one plausible explanation, that the world is, in reality, structured more or less as it appears to be, and, consequently, that the kinds distinguished in it are, for the most part, natural kinds.

(a) The hypothesis that the appearance of structure arises from our manner of perceiving or thinking about the world has no plausibility. It does not even begin to explain the structures that we find in chemistry, for example.

(b) The hypothesis that our mental processing systematically distorts reality in some way cannot be ruled out *a priori*, but in the absence of specifics, this is a gratuitous assumption, and the doubt is no more than a sceptical one.

3. If, in reality, the world is structured more or less as it is represented as being in the new scientific image, then the entities related in this structure must, for the most part, exist, be members of natural kinds, and have the causal powers that are supposed to be amongst their essential properties. Anti-realism about the essential properties of these entities would lead to anti-realism about the kinds, and therefore to anti-realism about the structure.

The realism of the new essentialism is a stronger form of scientific realism than most. It is, in fact, a form of scientific entity realism. But, at the same time, it relates only to those entities and properties that are described in the supposed causal structure of the world. It does not, for example, refer to the entities of the abstract model theories of thermodynamics, arithmetic, semantics or neo-classical economics. Consequently it is not committed to the reality of perfectly reversible heat engines, numbers, possible worlds or ideal economic agents.

Note also that the argument is not one that depends on a theory of reference or truth. The supposed basic epistemic rightness of the new scientific image of the world is *prima facie* compatible with any of a number of different theories of truth. It is, rather, an argument that seeks just to explain the plausible assumption of the basic epistemic rightness of this image. So, it gets down at once to the crucial question, which is: How is this picture of the multiply connected, highly interactive, hierarchical structure described in the new scientific image of the world to be explained? What gives rise to this image? The image is clearly a human construct. But it is an image that accommodates and explains just about everything that scientific investigation has revealed to us, and it excludes nearly every possibility that scientific investigation has excluded. This is what we should expect, if the scientific image is, for the most part, faithful to reality. It would be completely inexplicable, however, if the image were not a true reflection of the world.

5. CAUSAL POWERS AND AGENCY

Most philosophers today believe that the question of what causes what ultimately depends on what universal regularities hold. In Hume's original theory, A is said to be the cause of B, if and only if, A *precedes* B, A is *contiguous* (both spatially and temporally) with B, and events of the kind A are *regularly followed* by events of the kind B. This theory is called the 'regularity theory' of causation. On Hume's original theory, there are no genuine causal powers. An effect is not something that is somehow *necessitated* or *brought about* by its cause; it is just an event of a kind which happens to follow with universal regularity on events of the kind to which the cause belongs. Those who postulate that there are causal powers inherent in objects which are displayed in causal processes are accused of trading in obscurities. According to Hume, there are, really, no such things as causal powers, and causes do not necessitate their effects.

Hume's theory thus creates a problem for anyone who wishes to account for human agency. If Hume is right, then our conscious decision-making processes, and the actions which we say stem from them, must all be understood in terms of regularities, constant conjunctions, and the like, concerning which we, as conscious beings, can be nothing other than passive observers. But this is clearly not how they are in the manifest image of ourselves. We do not see ourselves as being in such a passive role vis-à-vis our own decisions and actions. Rather, we see ourselves as acting, and doing things for reasons. We see our processes of deliberation as ones that are thoroughly under our control, and which we can continue, suspend, or eventually act upon. Acceptance of a Humean theory of causation thus makes it very difficult for anyone to suggest a theory of human agency that is even remotely plausible.

More recent accounts of causation that belong to the Humean tradition are not, of course, all the same as Hume's. But most modern theories of causation that depend on counterfactual analyses, are not really very different from Hume's. They are all agreed that a case of causation is ultimately just an instance of a universal generalisation. They disagree with each other mainly about the nature and status of this generalisation. But more importantly, from our point of view, they all cast the agent into the role of spectator to his or her own decision-making processes. For an essentialist, however, agency is not a surprising or inexplicable phenomenon. On the contrary, everything is an agent of one kind or another. For everything has causal powers and everything is capable of exercising its powers.

Of course, the most elementary kinds of things all have fixed causal powers, i.e. their dispositional properties are all fixed by their essential natures. A copper atom, for example, has the same dispositional properties wherever or whenever it might occur. The same is true of a proton or an electron. They are things that belong to what might be termed 'fixed natural kinds'. Their distinguishing feature is that you cannot change any of their dispositional properties. They do what things of these kinds always do, and you cannot teach them any new tricks. There can be no question of a copper atom, for example, being disposed to behave in one way at one time, but in a different way at another time. Nothing with such variable powers could possibly be a copper atom.

For the fixed natural kinds there are universal laws of action. These laws are the causal laws. According to the new essentialism, all such laws are metaphysically necessary. They are metaphysically necessary, because things of these kinds have all of their dispositional properties essentially, and therefore could not possibly behave in ways other than as these properties dictate. The laws of chemical combination, for example, are necessary in this sense, as are the laws governing the behaviour of the fundamental particles. Things of these kinds must be disposed to behave as they do, because their identities as things of these kinds depend on their being so disposed.

As we ascend to more complex structures, we find that things that plausibly still belong to natural kinds have more variable dispositional properties. A piece of iron, for example, is plausibly a member of a natural kind, the members of which are all essentially crystalline structures of metallic iron. But pieces of iron can become fatigued, and hence brittle, or they may become magnetised, and hence acquire a capacity to attract other pieces of iron, generate electric currents, and so on. So pieces of iron may gain or lose causal powers, depending on their histories or circumstances. Moreover, some things evidently have the capacity to change the dispositional properties of other things. A bar magnet, for example, has the capacity to magnetise another piece of iron. A radioactive substance has the capacity to effect changes of various kinds in things in its vicinity, including changes of their dispositional properties.

In many cases, the induced changes in the causal powers of things do not constitute changes of *essential* nature. For the essential nature of a thing belonging to a natural kind is just the core set of its causal powers, capacities, structures, and so on, in virtue of which it is a member of the kind. A piece of iron is still a piece of iron, even if it becomes magnetised. In other cases, the changes induced in the powers of things by various kinds of agents do constitute changes of kind. The changes that occur to the working substance in an atomic bomb, for example, are transmutations that involve changes of essential nature. Any intrinsic properties or structures that a thing may either gain or lose while yet remaining a member of the kind are properties or structures that it has only accidentally. Those that it cannot lose are those that it has essentially.

At the next stage of organisational complexity, it seems that things may not only be made to acquire or lose dispositional properties by the exercise of extrinsic causal powers, as the case of a magnet magnetising another piece of iron well illustrates, they may also be made to acquire or lose dispositional properties by the exercise of certain other of *their own* causal powers, i.e. by the exercise of what may be termed 'meta-powers'. In general, an object may be said to have meta-powers if it has a reflective power to change some of its own causal powers. Inevitably, anything having meta-powers must appear to have higher-order powers, or powers of control. For any object that has such powers must have a capacity for self-direction.

It is plausible to suppose that human beings, and the members of all other advanced animal species, have acquired meta-powers of this kind in the process of evolution. Some of them may even have acquired meta-meta-powers, or powers of control of higher orders. If so, then human deliberation and action can fairly readily be explained. When someone acts to do something, they display a certain, perhaps very temporary, disposition. In at least some cases, this disposition results from an internal process of deliberation, a process that always involves the exercise of meta-powers. A deliberate action is not just an event of a kind that happens regularly to follow when intentional states of mind of a certain kind come into being. It is something that is done *by the agent* as a result of an intentional state of mind which is itself brought about by the agent, viz. by deliberation.

Thus, it seems that human beings not only have variable dispositional properties, as most complex systems have, but also meta-powers, i.e. *powers to change dispositional properties*. Other animals, no doubt, have similar meta-powers, but that such powers exist, and are exercised, seems quite evident from our own case. We exercise such meta-powers whenever we deliberate about what to do, and we call any action, which may result from this process, a deliberate act of will.

The new scientific essentialism thus promises to reshape the scientific image of man, as well as inanimate nature. It promises to do so in a way which will bring the scientific and manifest images of ourselves much closer together. For it deals with one aspect of the apparent conflict between them by providing a scientific image of human agency – an image which bears enough resemblance to its manifest counterpart for it to be taken seriously as telling us what human agency really is. If the new essentialism is accepted, then human agency could be accepted as the manifest image of actions brought about by people exercising their meta-causal powers, i.e. their powers of control.

If human agency is just the exercising of our meta-powers to alter our own dispositions to act in one way rather than another, then it follows that we must be able to monitor our mental processes, including our thinking, believing, desiring, and so on. That is, we must have a kind of second-order or meta-perception, or ability to know directly by experience something of what is going on in our own heads when we are engaged in any of these activities. The neuro-physiological basis for this meta-perception must be something like a meta-level neuro-physiological process, which scans the first-order processes involved in our various mental activities, including, it seems the activity of scanning. Consciousness, I would think, is just such a meta-level scanning process.

La Trobe University and the University of Melbourne

REFERENCES

Armstrong, D.M.: 1978, *Universals and Scientific Realism*, 2 Vols., Cambridge University Press, Cambridge.

Bealer, G.: 1987, 'The Philosophical Limits of Scientific Essentialism', in J. Tomberlin (ed.) *Philosophical Perspectives 1*, Ridgeway, Atascadero, pp. 289-365.

Bhaskar, R.: 1978, *A Realist Theory of Science*, Harvester Press, Hassocks, Sussex.

Bigelow, J.C.: 1999, 'Scientific Ellisianism', in H. Sankey (ed.), *Causation and Laws of Nature*, Kluwer, Dordrecht, pp. 56-76.

Burtt, E.A.: 1932, *Metaphysical Foundations of Modern Science*, Second Edition, Routledge and Kegan Paul, London.

Carroll, J.W.: 1994, *Laws of Nature*, Cambridge University Press, Cambridge.

Cartwright, N.: 1989, *Nature's Capacities and their Measurement*, Oxford University Press, Oxford.

Chalmers, A.F.: 1987, 'Bhaskar, Cartwright and Realism in Physics', *Methodology and Science* 20, 77-96.

Dretske, Fred I.: 1977, 'Laws of Nature', *Philosophy of Science* 44, 248-268.

Elder, C.L.: 1992, 'An Epistemological Defense of Realism about Necessity', *The Philosophical Quarterly* 42, 317-336.

Ellis, B.D.: 1990, *Truth and Objectivity*, Basil Blackwell, Oxford.

Ellis, B.D.: 2000, 'The New Essentialism and the Scientific Image of Mankind', *Epistemologia* 23, 101-122

Ellis, B.D.: 2001, *Scientific Essentialism*, Cambridge University Press, Cambridge.

Ellis, B.D.: Forthcoming, *The Philosophy of Nature: A Guide to the New Essentialism*, Acumen Press, Chesham, Bucks.

Fales, E.: 1990: *Causation and Universals*, Routledge and Kegan Paul, London.

Harré, R. and Madden, E.H.: 1975, *Causal Powers: A Theory of Natural Necessity*, Blackwell, Oxford.

Kripke, S.: 1972, 'Naming and Necessity', in D. Davidson and G. Harman (eds.) *Semantics of Natural Language*, Reidel, Dordrecht, pp. 252-355.

Lierse, C.E.: 1996, 'The Jerrybuilt House of Humeanism', in P.J. Riggs (ed.) *Natural Kinds, Laws of Nature and Scientific Methodology*, Kluwer, Dordrecht, pp. 29-48.

Martin, C.B.: 1993, 'Powers for Realists', in J. Bacon, K.K. Campbell and L. Reinhardt (eds.), *Ontology, Causality and Mind: Essays in Honour of D.M. Armstrong*, Cambridge University Press, Cambridge, pp. 175-94.

Maxwell, N.: 1968, 'Can there be Necessary Connections between Successive Events?', *British Journal for the Philosophy of Science* 19, 1-25.

Psillos, S.: 1999, *Scientific Realism: How Science Tracks the Truth*, Routledge, London and New York.

Putnam, H.: 1975, *Philosophical Papers*, Vol. 1: *Mathematics, Matter and Method*. Cambridge University Press, Cambridge.

Sellars, W.: 1963, *Science, Perception and Reality*, Routledge and Kegan Paul, London.

Shoemaker, S.: 1980, 'Causality and Properties', in P. van Inwagen, (ed.), *Time and Cause: Essays presented to Richard Taylor*, Reidel, Dordrecht, pp. 109-135.

Swoyer, Chris: 1982, 'The Nature of Natural Laws', *Australasian Journal of Philosophy* 60, 203-223.

Tooley, M.: 1977, 'The Nature of Laws', *Canadian Journal of Philosophy* 7, 667-698.

NOTES

[1]The new essentialism owes much to the work of Rom Harré and Edwin Madden (1975), who argued against Hume's theory of causation, and developed an alternative theory based the assumption that there are genuine causal powers in nature. But there have been many other contributors. In the late seventies and early eighties, Fred Dretske, Michael Tooley, David Armstrong, Chris Swoyer and John Carroll all developed strong alternatives to Hume's theory of the laws of nature, and their theories of natural necessity laid the foundations for the later essentialist ones. Sydney Shoemaker (1982) built on the earlier work of Rom Harré on causal powers to develop a thoroughly non-Humean theory of properties, which is a vital ingredient of the new essentialism. Roy Bhaskar (1978) developed a realist theory of science, which foreshadowed some of the later developments in the new essentialism. These philosophers are all distinguished by their realism about the objects, properties and processes described by science. That is, they are all scientific realists. They are also distinguished by their direct and speculative approach to metaphysics. For they consider the program of logical analysis that has dominated twentieth century philosophy to be irrelevant to the development of a sound metaphysics of nature. Metaphysics, these days, is unashamedly speculative. The new essentialism has evolved from these beginnings, and can now reasonably claim to be a comprehensive philosophy of nature. Many philosophers from around the world, including Sydney Shoemaker, Charles Martin, George Molnar, George Bealer, John Bigelow, Caroline Lierse, Evan Fales, Crawford Elder, Nicholas Maxwell, Nancy Cartwright and John Heil, have contributed in various ways to its development. So the new essentialism is not just a personal view, but an emerging metaphysical perspective that is the culmination of many different attempts to arrive at a satisfactory post-Humean philosophy of nature. Nevertheless, the perspective that I shall

present is my own, and I shall not attempt to differentiate it from others, or evaluate the other contributions that have been made to the general program.
[2]i.e. 'true' in the generic sense of this word. See Ellis (1990).
[3]Internal forces are really just external forces acting between the parts of things.

NOTES ON CONTRIBUTORS

LISA BORTOLOTTI is a PhD student in philosophy at the Research School of Social Sciences, Australian National University. She studied philosophy in Bologna, London and Oxford before starting her doctoral research. Her main interests include the relation between rationality and intentionality in the philosophy of mind, the epistemology of beliefs, mental illness and non-human animal cognition.

STEVE CLARKE is a research fellow in the Centre for Applied Philosophy and Public Ethics and a lecturer in philosophy at the School of Humanities and Social Sciences at Charles Sturt University. His work in philosophy of science includes papers in the *British Journal for the Philosophy of Science* and *International Studies in the Philosophy of Science* as well as a book, *Metaphysics and the Disunity of Scientific Knowledge*, (Ashgate, Aldershot 1998).

HERMAN DE REGT is assistant professor of philosophy of science and epistemology at Tiburg University. He is the author of *Representing the World by Scientific Theories. The Case for Scientific Realism* (Tilburg University Press, Tilburg 1995). In 2001 he was visiting scholar at the Department of History and Philosophy of Science, University of Cambridge. He has published articles on scientific realism, the rationality of belief, and American pragmatism.

BRIAN ELLIS is emeritus professor of philosophy at La Trobe University and professorial fellow in the Department of History and Philosophy of Science, University of Melbourne and is the author of a number of books and many articles on epistemology, metaphysics and philosophy of science. His latest books are *Scientific Essentialism* (Cambridge University Press, Cambridge 2001) and *The Philosophy of Nature* (Acumen Press, Chesham forthcoming).

MICHEL GHINS is currently a professor at the Université catholique de Louvain. He has taught at the Universities of Pittsburgh, Campinas (Brazil), Turin and the Catholic University of America. He has published *L'inertie et l'espace-temps absolu de Newton à Einstein: Une analyse philosophique* (Académie Royale de Belgique, Classe des lettres, Tome LXIX/2., 238, 1990) and a number of papers in history and philosophy of physics and general philosophy of science.

KEITH HUTCHISON completed a first degree in physics and mathematics, before shifting into the history and philosophy of science. He has published on: nineteenth century thermodynamics; the history of ideas in the Renaissance; probability; ancient astronomy; and time reversibility. He is particularly interested in the relations among science, philosophy, and political thought in early modern Europe.

HAROLD KINCAID is professor of philosophy and chair at the University of Alabama at Birmingham. He is the author of *Philosophical Foundations of the* Social Sciences (Cambridge University Press, Cambridge 1996), *Individualism and the Unity of Science* (Rowman and Littlefield, Lanham MD 1997), and numerous articles in the philosophy of science.

TIMOTHY D. LYONS is an Assistant Professor of Philosophy at Indiana University—Purdue University Indianapolis (IUPUI) where he teaches philosophy of science. He recently completed his PhD on scientific realism in the Department of History and Philosophy of Science, University of Melbourne.

ROBERT NOLA teaches philosophy at the University of Auckland. Most of his publications are in the area of philosophy of science. His most recent book is a collection, edited with Howard Sankey, *After Popper, Kuhn and Feyerabend: Recent Issues in Theories of Scientific Method*, (Kluwer, Dordrecht 1999).

JOHN WRIGHT teaches philosophy at the University of Newcastle, NSW, Australia. His main research interests are in metaphysics, particularly realism/anti-realism, and the philosophy of science, particularly science and rationality. He has published books on scientific rationality, and on realism/anti-realism, and also has a book forthcoming on the ethics of economic rationalism.

INDEX OF NAMES